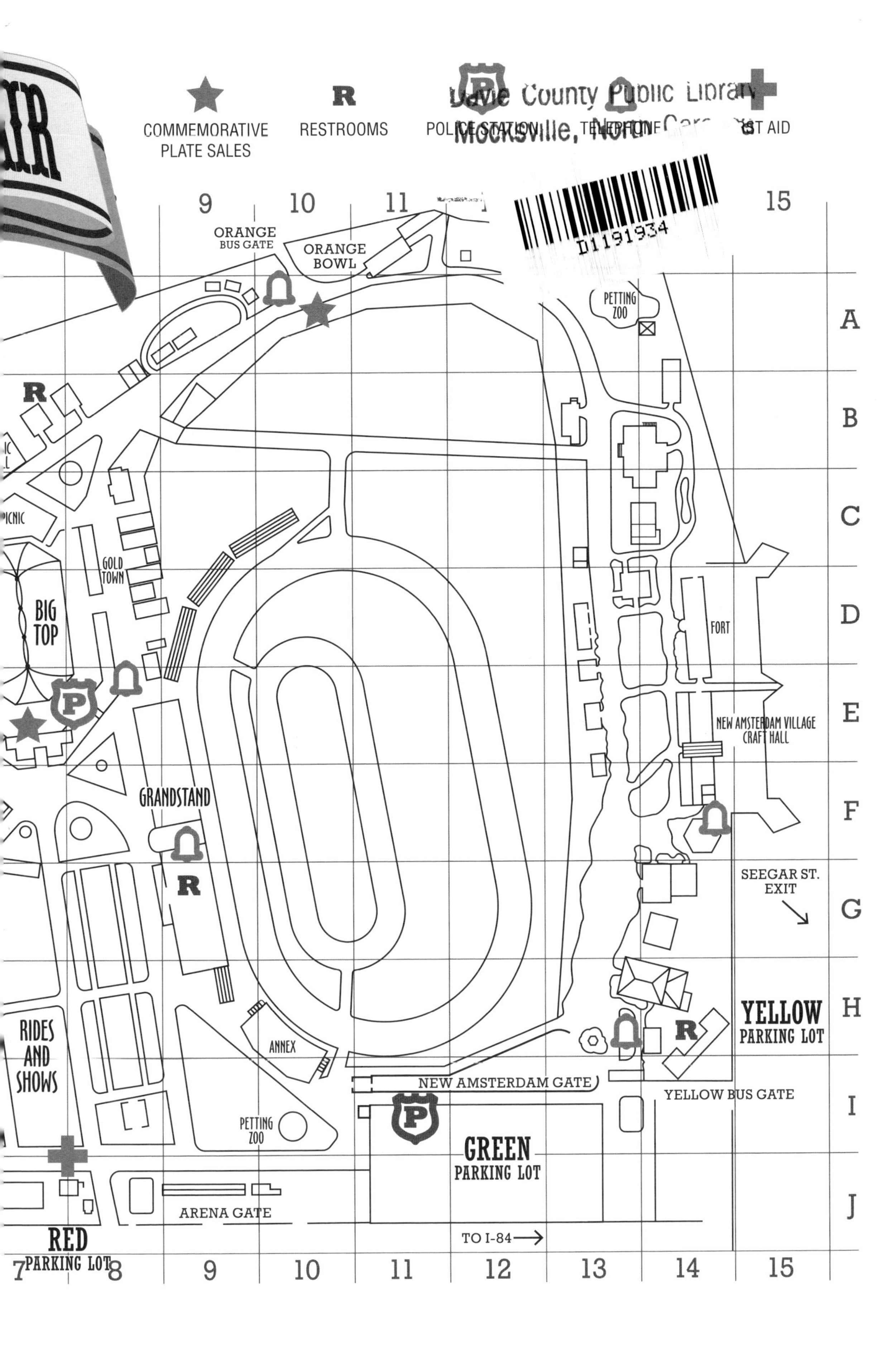
COMMEMORATIVE PLATE SALES
R
RESTROOMS
POLICE STATION
TELEPHONE
FIRST AID
9
10
11
15
A
B
C
D
E
F
G
H
I
J
ORANGE BUS GATE
ORANGE BOWL
PETTING ZOO
R
PICNIC
GOLD TOWN
BIG TOP
FORT
NEW AMSTERDAM VILLAGE CRAFT HALL
GRANDSTAND
R
SEEGAR ST. EXIT
YELLOW PARKING LOT
R
RIDES AND SHOWS
ANNEX
NEW AMSTERDAM GATE
YELLOW BUS GATE
PETTING ZOO
GREEN PARKING LOT
ARENA GATE
TO I-84
RED PARKING LOT
7
8
9
10
11
12
13
14
15

THE LIFE AND TIMES OF THE GREAT DANBURY STATE FAIR

by

GLADYS STETSON LEAHY
AND JOHN H. STETSON

EMERALD LAKE
BOOKS

Cover painting by Warren Baumgartner (1894-1963),
presented by *True* magazine to John Leahy on the 75th anniversary
of the Great Danbury State Fair.

Cover design and art by Gerber Studio
gerberstudio.com

Books published by Emerald Lake Books may be ordered through booksellers or by contacting:

Emerald Lake Books
Sherman, CT 06784

emeraldlakebooks.com

ISBN: 978-0-9965674-5-9 (hc)
ISBN: 978-0-9965674-6-6 (pb)
ISBN: 978-0-9965674-7-3 (epub)

Library of Congress Control Number: 2016950790

Printed in the United States of America

Dedication

To my wife, Carol, who accompanied me on
the crazy roller coaster ride that came with working
for John W. Leahy and the Danbury Fair.

CONTENTS

INTRODUCTION

The first part of this history of the Great Danbury State Fair, entitled "All This New Corn," was researched and written by my grandmother, Gladys Stetson Leahy, at the behest of her husband, successful entrepreneur and showman, John W. Leahy.

Knowing my grandmother the way that I do, I have a feeling the title contained a double-meaning. With her no-nonsense approach to life, the title may have reflected an opinion that the Fair and all its goings-on was a bit frivolous, but it also expresses the amazement of seeing something fresh come back year after year with careful tending and care.

Whatever her reason for titling it as she did, it was completed in 1956. After finding a "vanity" publisher who requested a $1,500 fee, which was a lot of money for her own frivolous venture, the manuscript was put away and the project was never completed.

After my grandmother's death shortly after the last Fair was held in 1981, I discovered the manuscript in the attic of their home. Being busy with closing out her estate, acquiring and operating our family's propane and fuel oil businesses, and raising my family, the book remained on my to-do list until 2015.

At that time, I decided to update the history to include the events that led to the sale of the Fair in 1982 and publish both parts as a whole. The second part of this book, "America's Most Unique Fair," is the embodiment of my effort to complete what my grandmother had begun.

Given that the Fair was not just an event, but an extension of our family, both parts share our personal recollections and firsthand narratives of the Fair we knew and loved so well.

JOHN H. STETSON

PART 1:
ALL THIS NEW CORN

by

Gladys Stetson Leahy

Containing among other things
A History of the Great Danbury Fair
From the Presidency of General Grant
To that of General Eisenhower

"From out of olde feldes, as men seith,
Cometh al this new corn fro yeer to yere."

The Parlement of Foules – *Geoffrey Chaucer*

1

LITTLE SHORT WORDS

My husband, John, has many novel and original ideas, which he attempts to translate into action as far and as fast as possible. Some of his undertakings have met with marked success while others have proved costly. All of them involve a lot of hard work and co-operation on the part of his assistants. So people like me, who like to jog along familiar well-worn paths, often try to dismiss sudden proposals of his that strike us as extreme or entailing unusual difficulties. This is hard to do.

One morning early in December, I stopped at our White Street office on my way shopping. Light-hearted and carefree, all I had intended to do was say a friendly "hello," pick up my check, and be off.

A wife who drops in and out of her husband's office at odd times cannot expect her welcome to be consistently enthusiastic, but John was glad to see me. Very pleased he seemed as he elbow-steered me toward an area of greater privacy.

"You know," he said, "before I went upstairs last night I was reading that book by a couple who ran a cucumber farm on Long Island. It was pretty good and I said to myself, 'If people like to read such stuff as that, why can't you and I write a book about the Fair?'"

"It was the wife who did the writing," I recalled.

"Yes, but she said, 'Without him this story would never have been told.'"

"That I am ready to believe."

"Now, a person doesn't have to be very smart to write a book like that," John persisted, "It's just a straight-forward story of what happened, told in a light, amusing way—all little short words."

He grinned as he began to help me off with my scarf.

"That's very encouraging," said I, clinging stubbornly to my coat, "but can't we wait until after the holidays? Couldn't we talk it over at home evenings? Christmas will be here in a few days and I'm late this year, as usual, getting started."

John is accustomed to some measure of resistance. Sometimes, in fact, it seems to act as a stimulant. On the other hand, he hates to have brakes applied as he roars enthusiastically along on the first fine rapture of a "new idea."

"What day is New Years?" he wanted to know.

"Sunday, January 1."

He pulled out his memo pad.

"We'll start it Monday, the second. Monday is a good day to start. You see this big envelope. I'll put all my clippings and any material I pick up right in here, and you can find it in the top drawer of the tall metal file in the corner. Now do you need some money?"

I admitted that a few extra dollars were exactly what it would take to cushion the shock that my nervous system had sustained, and then cheerfully went my way, not actually surprised or disturbed, but musing rather, that perhaps between us we could concoct some short history of the Fair.

This wasn't so bad. What if he had decided I ought to learn to play the clavichord? I have a friend whose husband, all unsolicited, brought home an Irish harp for her birthday.

Christmas came and went, and this conversation completely slipped my mind, but John has a memory system all his own. He keeps lots of little white notepads lying around the house and the office. To these he confides his plans and projects, each on a separate slip. These scraps of paper go into an envelope on which he has marked the date when he expects to give them detailed attention.

It works this way—John was in the office when he opened the envelope marked January 2 and found the following notations:

```
Pipe up new tank
Electricians wire pump
```

```
Paint over billboard
Take down Christmas decorations
Phone Mrs. Lotsatalk
Gladys-book
```

Whereupon he telephoned me.

"Hello," he purred, "Just called up to see how you're feeling this morning. Everything all right?"

"Oh yes, I'm fine. And you?"

"I'm fine, too. Are you going to be very busy today?"

"Not especially. Just housework and a few errands. Want me to go somewhere with you?"

"Well, not exactly. This is the day we were going to start the book about the Fair, remember?"

"Why, so it is," I promptly acknowledged. Then I tossed the ball back at him, thinking how pleased the radio show *Quick as a Flash* ought to be to get me on its program.

"Where do you want to work, John? All the information and newspaper clippings are down there in your office."

"That's all right. You don't need notes to begin with. You can make a start without those. Just a little background stuff about the Fair, and Danbury and maybe Connecticut in general."

"Perhaps I'd better begin with Connecticut and work the other way," I suggested.

"Well, the principal thing is to get going. Just write the first chapter today. I'll bring it down with me tomorrow and get it typed and we'll see what it looks like."

It looked like this:

A Little Background Stuff

Connecticut is called the "Nutmeg State," but the title was not pridefully self-bestowed.

There must be a lot of school children and others who believe that nutmegs are raised here, just as pine trees grown in Maine and peaches are a product of Georgia.

But, no! The truth is that early in our state's history, a few enterprising Yankee peddlers with time on their hands during the long winter evenings, whittled out of wood some specious imitations of nutmegs. When the frost had passed in the spring and the traveling got started again, they set out in their horse-drawn carts and peddled these nutmegs for the genuine article, along with their regular wares, to housewives far and near outside the state, and probably inside too, since these salesmen were not overscrupulous and had no notions of building up the territory or making repeat sales. All that came later.

As time went on, people about the country became suspicious of strangers from Connecticut and any native of that state was referred to as a "Nutmegger." Far from resentment, the recipients of the appellation must have thought this shady trick was pretty cute, for they adopted the title and they like to tell the story, as you can see.

Nowadays, peddlers have to have licenses and if they misrepresent the smallest item, some furious householder will call up the state police. As a result, wooden nutmegs are no longer sold, so far as I know. There are salesmen of certain securities, however, who ply their trade openly and make a good living, so that the most important difference between then and now is the classic one—the automobile has taken the place of the horse.

This indifference of Connecticut people to Puritanic standards is not in the tradition of a New England that looks to Boston as its leader in cultural and spiritual matters, but is probably due to the infiltration of New Yorkers, who do not retain a fast dollar long enough to ask it where it came from.

New York State is adjacent to Danbury's western boundary, and only sixty miles to the south is New York City, where many persons from this locality carry on businesses from sometime Monday morning until the Berkshire Express leaves the Grand Central on Friday afternoon.

If basketball players are being bribed in New York, we can scare up a couple who have been "approached." If torch sweaters go on sale in New York, the Danbury police quickly discover six or eight of them in our city.

Furthermore, in the summertime, cars pour out of the city into Connecticut all day long on Fridays, over that expensive and beautifully manicured highway, the Merritt Parkway. On Sundays, back they go at 50 to 60 mph, fender to fender and head to tail, like Hudson River shad in April. Their occupants love the quiet countryside, but as my Grandfather Cole used to say, "To make money you have to go where there is money," in spite of which bit of worldly wisdom, he kept right on raising apples on a Maine hillside till the day he died.

It is even getting hard in this locality to distinguish at first glance between the New Yorkers and the native-born. I can't do it, but there are subtle matters of grooming that furnish clues, I'm told, by friends who specialize in that sort of thing.

Many a New Yorker, whose house once stood within a convenient distance of the Connecticut border, has hired a heavy-hauler to move it down the road a piece into our state; from Brewster into Danbury, for example, or from North Salem to Ridgefield.

Such action was not taken to improve the view from the front porch, nor to ensure a bountiful water supply, but because up to now Connecticut has not levied a state income tax. However, indications are that our state government needs more money to pay more employees to inspect more things to protect the interests of more of us citizens. The protectors we have now are frequently individuals of high principle and standing in their communities, but they have families to support and most of them drive around in automobiles provided and maintained by the State. This runs into money and so, on the basis of present information, I doubt if our population will continue to be increased by this kind of immigration.

The citizens of Danbury are a decidedly chauvinistic people whose young are early instructed in these two essential facts:

#1 Danbury is a fine modern city offering unusual advantages and possessing superior public buildings and educational facilities.

#2 Every man, woman and child should wear a hat at all times and with every costume, excepting possibly pajamas.

A salesman who tries a house-to-house canvas without a hat is likely to lose confidence both in himself and his product.

One such unfortunate specimen knocked one hot summer afternoon at the door of a house where four ladies, all friends from way back, sat with their hats on playing bridge.

One of them rose briskly and confronted him through the screen. She glanced at his sample case, sized up his mission, and then before he had a chance to open his mouth...

"Young man," said she, "if you think you are going to sell anything in Danbury, calling around at people's houses without a hat on your head, you are very much mistaken. Don't you know that Danbury is the Hat Capital of the nation?"

He did after that, and the chances are he bought a hat in Danbury. At least he understood why, with the thermometer

John and Gladys Leahy on their honeymoon at Virginia Beach, October, 1935.

pushing 90°F, he had encountered at various front doors such a chilly atmosphere.

As for the Danbury Fair, I can say from personal experience that it is a cherished institution.

Twenty-two years ago,[1] John and I were married on Friday of Fair Week. John chose the date. It was none of my doing. He had been busy all summer building oil tanks in Norwalk and was anxious to get the ordeal over with before cold weather set in and his customers began calling up for oil deliveries. I am not native-born and so I supposed that he had just overlooked the Fair when he selected October ll.

Friday of Fair Week is Danbury Day, when old friends meet and those unfortunate exiles who have married out-of-town return to greet their fellows and to hoist a glass or some glasses to the days of yore.

Our wedding day was a perfect fair day. We were married at ten o'clock in St. Joseph's and when we emerged from the church into the mellow brilliance of sunshine and autumn leaves, there burst upon our eyes a sight that would have gladdened the heart of P.T. Barnum.

There were all of John's white tank trucks washed and polished by their drivers to a refulgence never before beheld in the annals of oil delivery, and decorated with crepe paper streamers. They stood lined up at the curb in order of size and importance, with two ten-wheelers at the head of the procession and three pickups bringing up the rear.

Their horns proclaimed us with strident, raucous uproar as we inched along up Main Street. This was a little too much for John's composure, which had just been sorely tried. He urged Charlie,

1 John and Gladys were married on October 11, 1935.

our cab driver, to speed up – then he tried to get him to detour via Elm Street, but it was no use. Charlie had other instructions.

In hilarious mood, the caravan followed us the whole length of Main and up West to Division, where they circled around the park and headed back with many a farewell blast.

John's face was very red for he felt sure people would think this was an advertising stunt. I was grateful and flattered and felt a surge of tender emotion toward the lightsome lads who had bothered to give us this noble send-off. To me, the parade was pure pleasure.

As we drove slowly past the fairgrounds on the way to New York, John watched the crowds lining up at the gates for tickets. I sensed his unspoken yearning.

"Too bad," I condoled, "we have to miss Danbury Day. We could have been married next week just as well."

"No," replied my brand-new husband. "I thought of that. Next week, the carpenters are coming to build a new loading rack for the bulk-yard on Pahquioque Avenue."

Since we had reservations for an afternoon sailing waiting for us at the pier of the Old Dominion Line, there was nothing for it but to continue on our course to Virginia Beach with our matron of honor and best man for company in case we didn't run into any fun-loving couples on the way. Like most of John's undertakings, the honeymoon was a success, but neither one of us, try as we may, can remember the name of that steamship.

"How do you like it?" I inquired of John as he finished a second careful reading.

"Some of the words are too long. Now that word 'chauvinistic'—I want people to be able to understand this."

"Very well. I shall try to limit myself to words of three syllables, but at times I may run over. I have to do it my own way."

"Then what you say about New Yorkers may offend them. We want to please everybody and have them like us."

"A fine thing!" I protested. "You ought to know better by this time. I have in mind a whole chapter calculated to displease certain obnoxious characters, the writing of which I was going to enjoy thoroughly."

"No, no." he said.

"John," I appealed to him, "do you think anybody is going to take this seriously?"

"You're always running into people who take things seriously. Now tomorrow we will write the second chapter."

"We will," I faltered, "out of what? Thin air?"

Thin Air

Two weeks after we were married, we moved hastily one evening into our present home on White Street, taking with us the contents of our respective small apartments.

John had purchased the house some months earlier when it happened to be for sale because it was handy to the office. The walking distance is five minutes or less, but of course we never walk.

We moved in the evening to save time. John and two helpers with a pickup truck transferred our belongings in a couple of hours. The efficiency of their performance was somewhat marred to my way of thinking by the loss en route of a box into which I had packed our table linens and some embroidered pillowcases. My best trousseau nightgown was on top, I recall.

They brought my sewing machine on the first load so that I could be shortening some curtains in order to give the front of the house, at least, a more lived-in appearance against the time when the sun would rise and the neighbors would awaken to find that we had stolen up on them. I never heard how they reacted to my efforts, but John approved them.

There really isn't much the matter with the house. It is sort of Dutch Colonial, white with green shutters and window boxes. Everybody says it is a "cute" little house. It is composed of six rooms, a sun-porch and a minimum of closet space. It does contain, thank goodness, a cellar and an attic, both somewhat difficult to access, into which we have managed over twenty years to cram the overflow of discards that we haven't sense enough to throw away.

John bought the place, as I have said, with an eye to convenience, but with both ears apparently obstructed.

The Leahys' first home at 205 White Street in Danbury, Connecticut.

Day and night, the flow of traffic past our front door is constant and tremendous. Early in the game, I took to using earplugs, but John can sleep through all but the most jarring of rear-end collisions. It takes the ambulance siren at close range to disturb him.

In the middle distance extend the tracks of the New York, New Haven and Hartford Railroad with plenty of long freights and a yard where switchers make up a train of eighty or more cars every night. The only complaint I have ever heard from John was uttered when the New Haven changed over from steam locomotives to the diesels, whose monotonous blast compared unfavorably in his ears to the old-fashioned train whistles.

"That's a hell of a noise," he would exclaim angrily as a fast freight stridently acknowledged one grade crossing after another, but he was only resenting being grown-up, and not the noise as such. He has lately even adopted the terminology of a neighbor of ours, who refers affectionately to the new diesel switcher as "Little Toot" because the sound it emits is less ear-splitting than that of the through locomotive, which passes as "Big Toot." "Big Blat" is a better name for it in my opinion.

In our little house by the side of the road, guests will often pause in mid-sentence to listen as our chrome-footed soap dishes go into their tap dance across the porcelain surface of the washstand upstairs.

"What's that?" they ask apprehensively as small cracks in the ceiling expand and new ones open up before their eyes.

"There's a certain amount of vibration here," John explains soberly. "Might be a vein of quicksand running under the house. I used to hear there was one in this locality."

For the first few years of our busy married life, I looked upon our residence as temporary shelter, a stopping place from which we would eventually emerge into a well-planned dream home. I postponed buying furniture until John expressed concern that the scarcity of our household goods might give the impression of impermanence, as if we were of two minds about continuing our life together. Even then, I carelessly picked up a few pieces that seemed good enough for as long as we were likely to need them.

I can't understand now what possessed me to think that we might ever get away. John never encouraged my flights of fancy. Every time I came across a promising site, it was too high or too swampy or too lonely, or it involved too much grass-cutting or was too difficult to access.

The most valid of his objections was voiced in the plaintive query, "Where would I go to lie down?"

It is true that the office is so near this haven of rest that, when the need arises, he can rush home for a cat-nap.

"Call me in twenty minutes," he will say as he stretches out luxuriously at the refreshing distance of twenty-five feet from White Street traffic.

I do and he rises invigorated. He can, moreover, in a pinch, dash home to change his clothes or shave—a great advantage, he feels. One thing I can say with conviction for life with John. It has been an interesting experience, but I must add, one for which I was signally unprepared.

Over the course of a few school-teaching years, I had always fancied myself as a homemaker. How blithely, I thought, I would go about my daily chores of cooking and dusting while I planned little surprise celebrations and social evenings for my pampered mate. What fun to make friends who would complete congenial foursomes for golf or bridge! How cozy for just the two of us, sitting by our own fireside, cat on a braided rug, to read and talk away the winter evenings.

The only part of that utopia that has materialized is the useless, necessary cat. Even now, when my erstwhile design for living seems somehow vaguely childish, he is an ideal companion. He has large yellow eyes, a smooth, dark blue coat and long legs that cause him to sit tall like Egyptian statues of cats. We called him "Farouk," partly for this reason and partly for other similarities to his namesake. He answers better to "Kitty."

Wintertime turned out to be our busy season with little leisure and phone calls at all hours.

To the oil business that possessed John when we were married and that possessed me for several years thereafter, he shortly proceeded to add a retail propane business, which as we say in our advertising "supplies gas to homes beyond the mains."

It was foolish for me to squander my energies on housework when I was needed to "chase slippers." A "slipper" in the parlance of our office is an inactive customer, one who used to buy, but whose record shows no recent order. My first job was to list such accounts, noting the unpaid balance, if any, and any equipment of ours the customer might have on loan. Later, when I began to post sales and chart deliveries, the deficiencies of my education became more apparent. A nodding acquaintance with bookkeeping and typing, for example, would have been helpful, but I had interest in my work, which John seemed to think would do, at least to start with.

There in the office I learned to hold my tongue. It wouldn't do to disagree with the boss publicly, and so I developed a sissy policy that calls for agreeing and cooperating with my husband.

I may hang back occasionally and I reserve the right to think my own thoughts, but outwardly I am polite. So is John, now I think of it, and we are each fairly tolerant of the other's eccentricities even in the privacy of our own home. No "blessings on the falling out" for us, as Lord Tennyson would have it. We are much too busy.

Very spineless, this sounds, I realize. Either that or we must have been particularly well-suited to each other. I can attest that the latter is certainly wide of the mark.

If commonality of background, tastes and interests are the ingredients of a successful marriage, John and I should have started running in opposite directions at first sight.

He is of Irish-German descent, rather a bizarre combination I still think, but one that has been the subject of considerable experimentation in these parts.

His Irish grandmother used to tell a story of how she left County Clare with her sister on a ship bound for New York with a stop-over in Boston. There the sister went ashore on a little sight-seeing excursion with somebody she met on shipboard. She stopped over too long and the ship sailed away to New York without her. The immigration authorities were more careless then than they are today, it would seem, and communication was not so highly developed. The young girls never heard from each other again.

After a brief sojourn in New York City, John's grandmother made her way to New Rochelle where she repaired her loss, to some degree, by finding and marrying another recent immigrant, whose name by coincidence was also Leahy. Her husband was a professional tooth-puller. He fashioned wooden teeth to take the place of those he extracted, so he must have had some mechanical ability, which he may well have transmitted to his grandson, John, who has that as well.

John and his Irish relatives are Catholics, but his father married into a German family whose members differ among themselves as much in religion as they do in sundry other ways. The German relatives are a jovial lot of non-eccentrics who follow their natural bent in normal and successful living.

I am more familiar with the quirks of my own people, who arrived in this country many boats earlier. I'm afraid I come from a long line of witch-swingers. I have no knowledge of any specific hag swung by my ancestors, but they had the temperament for it—at least those on my father's side had.

Once as a child of ten I wanted to go with another girl to a Sunday afternoon meeting of a children's group in the parish house of the Episcopal church. My grandmother would have let me go, but Grandpa put his foot down.

"I'd rather see her in her grave than an Episcopal," he announced angrily.

"In her grave" were his words, and I could see myself pathetically ensconced in a small white casket, wearing my best red and white silk dress. On my neck was my new locket and chain, and I bore a lily in my hand.

The matter was dropped right there, but for many years I thought I understood the meaning of the phrase, "a fate worse than death." It was to be anything other than a Methodist.

Now I was the apple of my grandfather's eye and I know he loved me dearly. I'm equally sure that he meant what he said and I hope he can't look down from heaven and see me now.

My mother's family were not such sticklers for doctrine, but more concerned with the moral aspect of religion. They were great believers in the beneficial properties of hard work. The highest praise of a stranger in their community was compressed into three little words, "He's a worker."

Satan could have little truck with a worker. My forebears would be all for John. Taking naps in the daytime or lying abed of a morning they condemned as weakening to both body and spirit, and they were not over-sympathetic with sickness, unless its symptoms were marked and severe. My own mother had recourse to a doctor only in childbirth and during her last illness.

Mainers are great on sins of omission. The most damning criticism of a housewife used to be the statement that she was handy with a can-opener. When I was a child, tearing ones clothes was a moral lapse indicating insufficient respect for the effort involved in obtaining them, and failure to mend them promptly was a mark of slackness. Not to return what one had borrowed was a disgrace. So was lack of judgment, which Mainers call "common sense."

I remember gratefully how my Maine grandmother used to guard my health by drying me out before the oven door of the kitchen stove after I had played in the wet snow, but then she would keep me indoors for the rest of the day. That wasn't so bad. I have been taught about colds and how to avoid them, but once I missed a long-awaited performance of "Uncle Tom's Cabin" by complaining of a bellyache and lying around the house most of a morning. Even though I was completely recovered by afternoon and it would be another year before Uncle Tom and his bloodhounds would be back, I had to stay home. It was like being punished for a misfortune.

Now John has common sense enough for an army and he works as if wound up, but his purpose in doing so is obscure. It is not from a sense of duty, I am thankful to state. More likely it's just for the hell of it. I find this attitude very relaxing to the compulsions of my upbringing, which doom me forever to be a mender of underwear and a turner of sheets. It does my heart good to see him pitch a pretty good pair of holey socks into the wastebasket, and the way he rips up his shirts with frayed collars for cleaning rags is a joy to behold.

I can spend a day in bed if I choose nursing an ingrown toenail or a tooth that I fondly, but foolishly, hope is not abscessing. At six o'clock, he will not consider it amiss for me to rise and dine and keep an evening engagement.

"Come on," he will say, "Get up and dress. You'll have a good time. I don't think there's much the matter with you. If there is, you can resume being sick tomorrow."

Then we both laugh. It is one of our private pleasantries.

Several years ago, we had bought tickets in February for a short southern cruise. Two days before we were to sail, I came down with what both the doctor and I diagnosed as bronchial pneumonia. Not even John, I felt sure, would allow me out of bed, but the day came and he aided and abetted me in getting packed and bundled into the warm car for the drive to New York.

The sun shone on my face as we bowled along the Henry Hudson Parkway.

"I feel better already," I croaked. "Only I hate to run out on the doctor like this."

"The warm air will do you good," John prescribed. "Anyway, there will be a doctor and a nurse on the ship and you'll get just as good care in your stateroom as you would at home."

I was afraid the French Line would not see eye-to-eye with him on this subject, but what with the sun and the car's heater and a few degrees of fever, I groggily decided that burial at sea might be a pleasantly cooling experience, if only I could stay alive long enough to enjoy it.

Once aboard, I crawled happily into bed and slept for a night and a day, whereupon the balmy southern breezes took over. I got up for dinner the next night and found it unnecessary to resume being sick.

Sometimes it is hard for me to recognize John's brand of common sense as such, but I have learned from him that the agreeable course may be both healthful and moral, and that was quite a revelation.

Until I knew John, I never cared much for sightseeing. All I know of its joys I learned from him. In a car, he is such a rubberneck that his eyes are everywhere but on the road. I am therefore always willing to drive and he is willing to let me. The first time I see a sight I am pretty good at enjoying it, but the next time we set out I want to see a new one. John can go on enjoying the same ones interminably.

Every Saturday after Thanksgiving, he boards the Commodore Vanderbilt at Harmon for Chicago to attend the annual convention of the National Association of Fairs. There, he follows a regular routine. He always attends the International Livestock Show, which takes place the same week. He visits the Brookfield and the Lincoln Park Zoo. He goes shopping for long periods at Marshall Field's without buying much, and he tours the city and environs on a sightseeing bus.

Soon after we were married, John had to go on a short business trip. As he packed hastily, I was eyeing his suitcase with some disapproval, for it looked as if the contents had been stirred with a spoon.

He must have misconstrued my attitude.

"Look here now," he admonished me. "You don't ever have to worry when I'm away. I shall be busy every minute of the day and all I ever take to bed with me is a pad and pencil. I haven't got time to watch you either," he added as an afterthought, "so let's get it understood now."

Not every bride might take kindly to this clear-cut declaration of policy, but what it lacked in sentiment I felt sure it more than made up in kindness of intention. It has made a fine working arrangement, so that now when he comes home from Chicago proudly bearing two large photographs of himself and Sally Rand in a dancing pose at the annual banquet of the Showman's League, I can smile and comment heartily, "What a gorgeous gown."

Sally Rand and John Leahy at the Sherman House in Chicago.

"She's a gorgeous dame," says John, who is a great admirer of the well-dressed woman and thoroughly enjoys being photographed with all the lively lovelies who show up at the Fair.

Sally, I know, is at the convention to arrange bookings for her act and, besequined as they are, those are her working clothes. My working clothes are plainer, if somewhat more ample, but I am solidly booked for all the years ahead and a great satisfaction it is.

"Not much there about the Fair," was John's comment, although he seemed pleased with my testimonial to his character.

"You are the one who grew up in Danbury," I retorted. "Why don't you write your own reminiscences?"

"After dinner tonight I'll tell you the things I remember about Fair Week when I was a boy. I remember quite a lot," he mused, "it may take more than one evening. After I run dry you can talk with Jarvis.[2] He can tell you what was what, and after that—why, you've been around here for twenty-odd years, you must have a few ideas of your own."

I had a few all right, but they wouldn't bear mentioning.

2 C. Irving Jarvis, longtime Assistant General Manager of the Fair, was in charge of seeing John's ideas through to fruition.

Boy's Eye View

John loves people and the more the merrier. For the twenty-five years I have known him, he has shown a marked predilection for parades, circuses and carnivals, auctions and town meetings, fires and funerals, and in fact for any goings-on where his fellow creatures may be gathered together.

In the summer of 1949, nearby Bridgeport, where P.T. Barnum lived in Oriental magnificence during his heyday and which was the winter quarters for "The Greatest on Earth," held a Barnum Centennial to celebrate the organizing of "Barnum's Great Asiatic Caravan, Museum and Menagerie" in 1849.

Operating expenses were raised by public subscription, principally among the city's merchants, who gave liberally from civic pride and in the hope that business would be brisk as the crowds flocked into town. There were speeches, parades and a pageant in which young and old took part. In the evening, Seaside Park, many acres of which were given by Barnum to the city, was the scene of free fireworks. Everyone who could stand the heat went to the freight yards to see the circus unload, for its arrival had been planned to coincide with the jubilee.

There have been rumors that the centennial was not a financial success from the viewpoint of those who contributed to this promotion. If this is true, Bridgeport failed to profit by the precepts and example of her foster child who is believed to have said, "There's a sucker born every minute." Be that as it may, the celebration has become an annual event and is now said to be breaking even, which is still a long way from appropriate recognition of Barnum's talents.

It is, however, just the sort of thing John would crawl miles to

A collection of Mardi Gras heads parading along at the Danbury Fair. *James J. Tessasa*

witness and, in 1952, he participated by entering our Danbury Fair bandwagon in the parade, filling its seats with big boys wearing Mardi Gras heads. Sometime afterward, to his surprise and pleasure, he received by mail from Bridgeport a box containing a shield-shaped trophy, which makes a fine conversation piece for the wall of our reception room at the fairgrounds. "Most Comical" is the inscription.

John Leahy as a boy, enjoying his dog.

No St. Patrick's Day parade on Fifth Avenue ever roused John to such enthusiasm as the Barnum Centennial. It carried him back to the happy mornings over forty years ago when he had been up and going at five o'clock to meet the circus train.

There is always something that bends the twig. One glimpse of romance and a boy runs away to sea or a girl is stage-struck. It must have been the circus that inclined John to select as his personal hero our favorite native son, the famous P.T. Barnum. Emulation at the age of eight, which took the form of holding pet shows in the backyard, is one thing. Reverence at 54 for the same idol will bear looking into.

Was the profitable exhibition of Tom Thumb and the Cardiff Giant the measure of Barnum's greatness? I wondered.

It was enlightening to discover that, while many of his exploits have become legendary, his achievements as a solid citizen are less well-recollected. A giant in the nutmeg tradition was Phineas Taylor Barnum, who was born in 1810 in the adjoining town of Bethel when that village was a portion of Danbury still, which it had remained until 1855. Each community claims him as its own and reasonably so I should think without any need for further debate or research.

His name and those of his prodigies are in use today to glamorize many a business enterprise and city street, but Barnum left his mark on more than the landscape in these western reaches of Connecticut.

In addition to carrying on the activities for which he is more widely known, he was the Mayor of Bridgeport, President of the Pequonnock National Bank of Bridgeport, President of the Bridgeport Hospital and of the Bridgeport Water Company, and a member of the state legislature for four terms.

The genius and originality of his advertising methods started, for better or worse, a new trend in that whole field, making him the acknowledged father of modern publicity. In his early days, he published his own newspaper and later in life found time in his odd moments at home to write a book for boys entitled "The Adventures of Lion Jack," as well as to complete a lengthy autobiography, which sold half a million copies.

Along the way, he contrived to entertain his friends and associates often and with profuse liberality at his palatial Bridgeport home, Iranistan. One close friend and a frequent visitor was Mark Twain, who might well have repaid his host's hospitality by writing a piece about the circus, but who, in spite of repeated urgings, never did.

For several years, Barnum was also President of the Fairfield County Agricultural Society and managed the fairs it held. This organization was established as far back as 1821 "for the improvement and encouragement of agriculture, domestic manufactures, industry and economy, and the holding of annual cattle shows and fairs in some town in Fairfield County during the month of October."[3] The fair was to be held in whichever town offered the largest cash inducement.

"It was held in Danbury in the years 1857, 1858, 1860 and 1863, in Bridgeport in 1861, in Stamford in 1853 and 1854. In all other years from 1855 to 1866, it was held in Norwalk. In 1867, the society purchased and fitted up grounds at Norwalk where its annual fair was held until 1888, when it was discontinued and the grounds sold."[4]

Thus in his veneration of Barnum, John was unwittingly paying homage not only to a great Nutmegger, but to the resident and promoter of the agricultural society that must be regarded as the progenitor of the Danbury Fair.

Since Barnum went to his reward in 1891, John was destined never to know his hero, but by the turn of the century the

3 *History of Danbury*, J. M. Bailey

4 Ibid.

Danbury Fair was in its prime and ideally designed to sustain a boy's interest in showmanship.

John sticks to it that he was 21 before he slept in a bed during Fair Week. This sounds like an exaggerated claim, but it may be true.

It seems that his mother had six sisters and three brothers who were scattered by reason of marriage and similar exigencies all over Westchester, eastern New Jersey and the Bronx, and that most of these contrived to take their vacations the week of the Danbury Fair. People still do this and for the same reasons.

For many years, the first week in October has borne a good reputation locally for weather. In fact, after the September gales and occasional hurricanes, there is usually a period of sunny days and cloudless skies that we call "Fair" weather.

The nights, however, are likely to be chilly, and regularly each year there arose in the home of John's parents the problem of whether or not to set up the parlor stove.

This efficient heater was known as "the Round Oak." Almost every family had one. It was truly useful and the makers had meant it to be decorative as well, for it wore two encircling frills of nickel-plated cast iron, a narrow one at its shoulder line and a wider one about the abdomen. It also bore a crown of the same, surmounted by a handsome bronze Indian, tomahawk in hand.

The stove, which had rested dismantled in the woodshed throughout the summer months, would have to be set up soon anyway, but there were two good reasons for postponement.

The first objection was on aesthetic grounds. Since the stove-pipe had to travel the length of the room to reach the chimney, and the lengths of pipe and elbows were suspended by wires from hooks in the ceiling, John's mother thought her parlor appeared to better advantage without these practical touches.

Secondly, the absence of the stove made for more sleeping space, for in its place the Larkin soap Morris chair[5] could be extended to accommodate one more visiting relative.

This burning question was settled in a hurry one Wednesday night when the weather turned suddenly cold. The guests turned to, bringing in the pieces of Round Oak and sections of stovepipe, and by dint of earnest striving completed the assembly, pipes, wires,

5 Purchases of Larkin Soap were rewarded with premiums, which could include chairs, desks and sundry other items. The Morris chair is a style of chair with a reclining back that could also be used as a bed.

hooks and all by 11:15 p.m. They were well-blackened with soot by that time, one uncle had jammed a finger, and a few harsh words had been spoken, but after a fire had been kindled and some hot toddies consumed, they became once more a happy and united family.

In homes all over town, hospitality was the watchword. For days in advance, women bustled about the house, husbands raked and spruced up outside with touches of fresh paint here and there, and October's breath was sweet with scents of pies baking and leaves burning.

It is a great thing for a whole community, be it large or small, to have a traditional annual festival in anticipation of which the womenfolk can plan costumes, trim hats and save recipes for months ahead, while the men groom livestock for blue ribbons and cash prizes. The Danbury Fair has always combined the best features of a race meet, a stock show, a Mardi Gras and Old Home Week, just as the first light frost has touched the maples and elms, the beech, birch and sumac, as if the countryside is simultaneously running a pageant of its own.

Born and raised on Balmforth Avenue hard by the New York, New Haven and Hartford railroad station, in the days when everything moved by rail, John had a box seat for seeing the sights as they proceeded from train to platform.

Many of the shows and rides with their equipment and personnel came from New York City, but excursion trains from Bridgeport, New Haven, Waterbury, Stamford and Norwalk pulled into the White Street station as well. From this point, a shuttle train operated every quarter hour over the two miles between Danbury and the fairgrounds.

Furthermore, Balmforth Avenue was the main artery for New Fairfield traffic heading to the Fair. On a Saturday, came the dairy cattle and teams of oxen, the draft horses pulling carts full of sheep and goats, poultry and rabbits, the wagons and buggies from the Hatch Carriage Factory—all passing the house in delightful preview.

John would go with his father to meet the New York Express that chugged into Danbury at 8:15 p.m. on Sunday evening, laden with home-comers and out-of-towners, equally irrepressible, commingling with giants, midgets, glass blowers, sword-swallowers and sideshow freaks. It always took them some time to pick out their relatives and escort them home, where a great unpacking took place and sleeping arrangements were gone over lightly.

Get your ticket! It's only 10¢ to see
Mickey Mouse's Smallest Living Circus on Earth, 1931.
Or check out Singer's Congress of Human Freaks instead, with
Madam Naomi, Mentalist Supreme, and The Lobster Boy, 1933.
Frank Baisley

Very little boys fit pretty well into a couple of armchairs pulled up to face each other. The next size can be accommodated if an ironing board is used as a sort of extender to bridge a gap between the chair seats, and a twelve-year-old likes nothing better than a mattress on the floor. If he does object, he makes no moan. Vacationing guests are often slightly conscience-stricken at putting the son of the house out of his bed, especially aunts, and loose change frequently makes its way from their pockets to the poor boy who was ousted.

A nickel was important money in those days. It would buy a hot dog, a bottle of soda pop, a red candy apple, an ice cream cone, a merry-go-round ride, admission to many of the side shows or three hoops to ring a cane.

In spite of the magnificent purchasing power of small change, quite a lot of townspeople didn't have enough of it to enjoy the week properly. This was not always due to circumstances beyond their control either, for Danburians are a fun-loving lot.

John tells a story of a certain neighbor family who took a drastic step one year in order to go to the Fair. They had run out of spending money so that even the 75¢ admission price was beyond their means, not an alarming situation ordinarily, but one that posed a bit of a problem.

The weather was mild that fall and there was the prospect of work in the hat shops a little later. They knew, moreover, where they could get $15 for the Glenwood kitchen range. They quickly hunted up the prospect, closed the sale, took the $15 and had fun all that week.

"What came of this piece of recklessness?" I asked John, hoping to hear that Providence had rescued the improvident and that such fine spirits had not been chastened, but he seemed to consider the question irrelevant.

"Why, after the Fair was over, they didn't have anything to cook their meals on, the half-wits," replied my realist husband.

And that was all I could find out about that!

Once it starts, however, John's flow of reminiscence usually swells to a broader tide. Those were the happy days. A question or two to show interest in the subject and I can sit back.

"When I was sixteen," he recalled, "I worked for Randall Harrington."

"You mean Mr. Harrington, the hat manufacturer?" I queried. Mr. Harrington is strictly gentry.

"The same. It was while he was living on Park Avenue."

"In Danbury?"

"Of course, in Danbury where else would he live?"

"Never mind me. The Harrington's are a fine old family."

"Well, Randall had a fine fiery temper. I can't remember how many times he fired me and then sent for me to come back—half a dozen, at least. After the first couple of times, I could tell when he was getting ready to blow off steam and if there was time I would make myself scarce before he popped."

"And you kept coming back for more? You really liked the job?"

"I needed the job. Now if you want to stop asking questions and listen, I'll tell you about Randall's rooster."

And he did.

It seems that Mr. Harrington was grooming a pen full of white Wyandottes[6] for first prize in the poultry show 1912.

About the 1st of September, Snowball, the big boss of the Wyandotte seraglio,[7] suddenly began to molt some of his tail feathers. He was still blessed with tail enough to rejoice the average rooster, but Mr. Harrington preferred to take no chances. He

6 A breed of chicken that are a docile, dual-purpose breed kept for their brown eggs and for meat.

7 A harem's living quarters.

rushed a letter enclosing $25 cash with the order for a replacement to some rooster grower in Ohio.

A period of anxiety ensued, then came an acknowledgement of the order. A fine cock would be shipped in ample time to pinch hit for Snowball. It would arrive the week before the Fair opened.

More days passed, again the tension mounted. Telegrams were exchanged between Danbury and Ohio. The express office was alerted and standing by to announce the arrival of the prize bird.

In late afternoon the Saturday before the Fair, when hope had reached its lowest ebb, the express office phoned to say that Mr. Harrington's shipment had arrived.

John was dispatched in the two-cylinder red Maxwell runabout. On the station platform, he found a crate in which apathetically sat a very tired-looking rooster. His comb was nicked and bloody, his plumage dirty and tattered. The tail feathers he had were not worth counting.

John hunted up the express agent and they stood together looking into the crate.

"What do you think happened to him," John asked.

"Dunno. Looks like he got run over. Perhaps he was too long on the way. Wanna file a claim?"

John distastefully took the rooster aboard and drove the Maxwell slowly back to the stable, where his boss was waiting impatiently.

"How is he?" Mr. Harrington demanded.

"You look at him," replied John heaving the crate to the barn floor.

Then seeing that hope deferred had rendered his employer touchy, he slipped out the back door and went home.

The next morning, he thought it was safe to report for work.

He found Mr. Harrington and the gardener engaged in loading a pen of poultry on a small truck.

"Get hold of that pen," Mr. Harrington shouted and John got hold fast, omitting to ask any questions on the way to the fairgrounds. When the chickens were being transferred to the exhibition pens, he saw it was Snowball that had been chosen for the competition, and never had his tail feathers appeared more luxuriant.

Later in the day after the judges had made their rounds John peeked in again. There strutted Snowball, still cock-of-the-walk, with a blue ribbon tacked to his cage, before which his beaming

> owner was holding court, explaining to an admiring coterie[8] the secret of raising Wyandottes.
>
> Surprisingly enough, however, the molting process seemed to have set in again, as evidenced by long white tail feathers all over the floor of the cage. It looked mightily as if Snowball was coming unstuck.

John says that the local entertainment attractions during his boyhood were drinking, baseball, the Taylor Opera House, Lake Kenosia and the Danbury Fair, in that order.

Of these, drinking and baseball have endured with some modifications in ground rules and conventions, but with a more or less specialized following.

The Taylor Opera House, a large wooden structure that stood at the corner of Main and West Streets where the Pershing Building is now located, once saw year-round service as theatre, concert hall and auditorium, but it burned to the ground one cold winter's night and was never rebuilt.

Gone, too, and almost forgotten, is the park at Lake Kenosia.

Lots of young people who have lived here all their lives have no idea that this pleasant little pond just west of the fairgrounds was once a popular resort complete with hotel, outdoor theatre, dancing pavilion and bathing beach.

I certainly had no inkling of it until I went to a party with John. As the evening wore on and inhibitions wore off, the event developed into an amateur talent show. At the close of a song and dance act by our hostess, John, pad and pencil in hand, scurried to congratulate her.

"Great!" he exclaimed as the applause slackened, "you've got the stuff, Betty. I'll book you for six weeks at Lake Kenosia."

Since most of the natives present were old enough to have known the lake in its heyday, this piece of waggery struck them as excruciatingly funny.

In 1895, many years before the Connecticut Light and Power Company flooded the valley to the north of Danbury to make what is now Candlewood Lake, an older company, The Danbury Horse Railway, undertook the development of Lake Kenosia as an amusement park. It began by purchasing a few acres at one end of the lake and laying tracks for the new electric trolley cars a distance of about three miles from the center of town. These tracks paralleled the south side of the fairgrounds.

On the shorefront, a pavilion and summer theatre were built, which were leased for operation since the company's interest was in the revenue from fares on its new line.

---

8 An exclusive community or clique, who has similar interests or tastes.

The investment proved profitable. Trolley cars making the run from Wooster Square did a steady weekday business and were jam-packed on Sundays. There was swimming and fishing with picnics in the grove. There were boats for hire, a big rickety roller coaster and a merry-go-round with brass rings to catch for a free ride. The popcorn business was terrific and a bar in the pavilion did pretty well, too.

Patrons of the theatre eagerly awaited the weekly offering of the resident stock company. Pre-curtain time was murmurous, John tells me, with conjecture among the ladies touching which of her five gowns the heroine would elect for that performance, while the first appearance of the villain drew prolonged appropriate hisses from the whole audience. The private lives of the actors became as interesting to their following as anything they did on the stage. In that pre-cocktail era, they were depended upon by many a hostess to stimulate the flow of table talk.

Visitors enjoy picnicking, boating and other activities at Lake Kenosia. *Danbury Museum and Historical Society.*

In 1915 came the blight of World War I, which shriveled the park's profits. Its popularity was further diminished by the development of the automobile to such a point of perfection that it was no longer necessary to include a mechanic when readying the family car for the rough twenty-mile trip to Norwalk.

Finally, Henry Ford came out with his $350 Model T and in five more years most of Lake Kenosia's patrons had converted themselves into Sunday drivers.

The park's decease in 1922 left the trolley line that ran past the fairgrounds available during Fair Week, and the Fair, in another more important way, became its beneficiary.

Irving Jarvis, the present assistant manager of the Danbury Fair, grew up in the amusement business at Kenosia. His father, William Jarvis, and his uncle, Leo Lesieur, were the lessees and operators of the park in its heyday. There young Irv alternately rented boats, sold tickets, ran the merry-go-round and swept up. He learned to repair electrical circuits and to tinker with sound systems. He also found out a great deal about human nature.

The Jarvis family owned property adjoining the park and continued to live there after the business closed. William Jarvis became associated with the Fair as assistant to the vice president, C. S. McLean, who was in charge of selling concession space and booking acts.

Irv worked with his father accumulating the experience and know-how that have made him a crack man among fair men. Not long after his father's death, he joined the management of the Danbury Fair on a full-time basis.

Last on the list local entertainment attractions, the Danbury Fair has survived four great fires and two World Wars. Constant, if not changeless, it has held its ground, improved its facilities and extended its run to nine days.[9] Its durability is no accident.

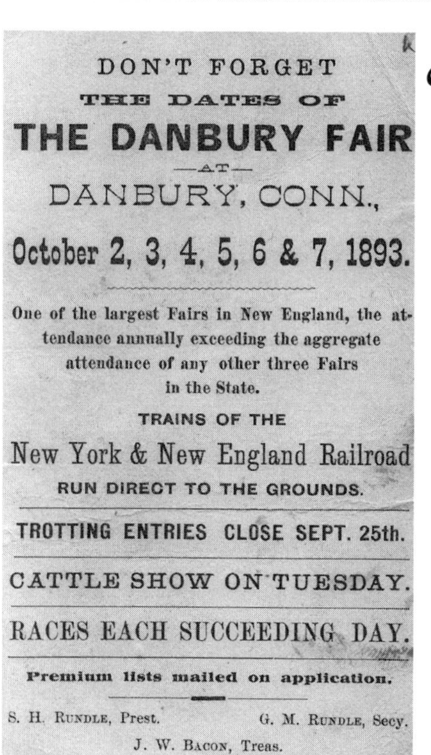
DON'T FORGET
THE DATES OF
THE DANBURY FAIR
—AT—
DANBURY, CONN.,
October 2, 3, 4, 5, 6 & 7, 1893.

One of the largest Fairs in New England, the attendance annually exceeding the aggregate attendance of any other three Fairs in the State.

TRAINS OF THE
New York & New England Railroad
RUN DIRECT TO THE GROUNDS.

TROTTING ENTRIES CLOSE SEPT. 25th.

CATTLE SHOW ON TUESDAY.

RACES EACH SUCCEEDING DAY.

Premium lists mailed on application.

S. H. Rundle, Prest. G. M. Rundle, Secy.
J. W. Bacon, Treas.

Down to Business

"Have you started on the history stuff yet?" John demanded as he opened the front door one day in February and found me with books and writing materials spread out north, south, east and west.

Actually, I was deep in *The Cruel Sea* and had been fighting submarines for an hour or two, but I am gradually sloughing off that Puritanic sense of guilt that used to make me drop a book at the sound of an approaching footfall and so I didn't drop it. I just let it mingle gently with the debris about me and tried to look as unconcerned as a chameleon on a rubber plant.

"No, but I have some books here from the library," I replied truthfully, "and I've sharpened a lot of pencils."

"Here's something I saved for you."

I glanced at the clipping.

9 When the federal government declared, in the early 1970s, that Columbus Day would be celebrated on the second Monday of October, the Fair was extended to ten days to include that holiday.

John is devoted to Americana—and ice cream. "April 23, 1951," it read, "marked the 100th anniversary of the ice cream soda."

"That will help a lot," I commented dourly. "And had you heard that the earmuff was invented by Chester Greenwood of Farmington, Maine, in 1877?"

"All right, smarty, but you'd better start producing."

"I always despised history." I said—to myself.

The truth was that my present assignment reminded me of a defeat I once sustained by a history instructor who prefaced an examination with the encouraging announcement that it would contain only one question.

The question as advertised read: "Outline the principal events that took place during the Hundred Years' War and fill in the out1ine."

"That lazy joker," I said to myself. "History teachers have been working that one with variations since before the Hundred Years' War."

A persistent rumor that this individual never corrected papers anyway, but rather threw them upstairs, grading them in the order in which they came down, moved me. In a shameless display of light-mindedness, I just filled the pages of my blue book with nonsense rhymes for which, by contrast with history, I had a remarkable memory. Starting off with "Beware the Jabberwock, my son,"[10] I ended with "Waste,"[11] a favorite of my girlhood, to wit:

Our governess—would you believe it?
Drowned herself on Christmas Eve.
This was a waste, as, anyway,
It would have been a holiday.

As it turned out, he must have read some of the papers. At least he came across mine, and we had to have a long serious talk before he would again consider me as a possible contender for history honors. In the course of the interview, he described my attitude with the word "intransigent," which, upon consultation with a dictionary, I interpreted to indicate that I have great perseverance.

That quality being my strong suit, I abandoned *The Cruel Sea* for the time being and decided to make an honest effort.

As I painfully progressed with my research, I discovered that the custom of holding annual fairs did not originate in Danbury. Neither was it devised by the Russians, who might conceivably claim another first in this field of endeavor, since they used to run a great and famous one at Nizhny-Novgorod, a city located in the very heart of European Russia at the junction of the Oka and Volga Rivers. Well over 100 years ago, this world-famous annual fair reported sales of merchandise in the amount of 150 million rubles, which was a respectable amount

10 Lewis Carroll, *Alice's Adventures in Wonderland*

11 Harry Graham, aka Col. D. Streamer, *Waste*

of rubles, equivalent to $112 million at the time. But a lot of water has flowed over the dam at Nizhny-Novgorod since those halcyon days. *Billboard* reports that attendance has fallen and sales have dropped off in recent years. The name of the city is another casualty. I don't suppose the Russians may be said to rechristen a city, but they now call that one Gorky,[12] which certainly is easier to pronounce, if less intriguing.

Actually, fairs seem to have been going on as long as history has been written. I shouldn't be at all surprised if Stone Age men picked up their axes, hammers and mallets and got together every fall to compare progress and steal each other's designs.

By Biblical times, fairs were in full swing. Ezekiel repeatedly mentioned the fairs of Tyre in enumerating the riches and commerce of that port before the wrath of God descended upon it.

Among the Romans, holy days were called "feria," from which our word "fair" is derived. At these religious festivals, great numbers of people gathered for worship, feasting, games and general hilarity. In rural areas where there were no markets, the country folk grew into the habit of bringing and exposing for sale various commodities and products of handicraft. The result was that bartering among the people became a regular adjunct of the holy program.

This routine was continued by the Italians of the middle ages with an important result. The seaports of Italy exerted a powerful influence on commerce. Vessels from many distant countries crowded the port of Venice for its great fairs. The need for a medium of exchange among these different peoples prompted the leading schemers of the day to think up a system of credits to take the place of money and so led to the invention of international banking.

For many centuries, fairs were related to commerce, buyers from rural areas would come to centers of population for an annual shopping spree. Indeed, European fairs are still conducted as businesses.

As country communities grew into towns and villages where it was possible to buy in regular markets, people were less eager to travel long distances to visit big city fairs. Small fairs nearer to home became characteristic of rural areas, particularly those devoid of railroads. The accent was still on buying and selling, but trading with local merchants at their leisure made buyers more quality-conscious.

Probably soon after the first young farmer was persuaded to journey to a fair over the country lanes with his wife behind him on pillion, comparative shopping was instituted. After that, pride of ownership and craftsmanship led to the display of goods not primarily intended for sale. Since farm animals and products were the principal interests of country folk, emphasis at country fairs gradually shifted from business to agriculture.

---

12 In 1991, this was changed back to Nizhny-Novgorod.

This country had no fairs in its early colonial days, partly due to the sparseness of population and partly because families had all they could do to produce for their own consumption.

There was a society formed in New Haven in 1803 for the promotion of agriculture and in 1804 a fair was held in the District of Columbia. Yet it took a retired merchant from Pittsfield, Massachusetts, to inaugurate our present system of agricultural fairs and cattle shows.

In 1807, a certain Elkanah Watson tied two merino sheep under an elm tree on the village green in Pittsfield and stood by to observe public reaction.

Response was so prompt and the gathering so numerous that Elkanah, who was a man of vision, decided to try to interest the farmers of Berkshire County in holding annual exhibitions of improved breeds of cattle and superior products of the soil in order to show what might be accomplished by proper culture. His undertakings led to the establishment of the earliest American agricultural fairs in New England and in upper New York state.

The foregoing ought to just about take care of fairs in general, I decided, and pave the way for the Danbury Fair in particular.

But then what?

Even from John's deep well of anecdotes little could be dredged up to my purpose for the reason that there were twenty-five years during which he and the Danbury Fair did not run concurrently and probably five more before his powers of observation really began to take hold.

Those thirty years were a problem, since I could find no orderly chronicle of the Fair beyond the account of its founding in the *History of Danbury, Conn., 1864-1896* by J. M. Bailey, known as "The Danbury News Man." As I followed Mr. Bailey's detailed account of the period in question, I did come to the conclusion that in the 19th century printing the news rivaled hatting as a principal industry.

Instead of the one *Danbury News-Times*, which now tops off our after-dinner hour, Mr. Bailey listed seven separate papers that, over those years, were launched upon careers of brief duration, along with four others that were still above ground in his time. Among the survivors were two newspapers, *The Danbury Dispatch* and Mr. Bailey's own *The Danbury News*.

In the columns of the *News* and the *Dispatch*, I felt pretty sure the Fair must have been reported and liberally commented upon.

There was a catch to that, of course. Few copies from so long ago exist intact. Most of them are in the old files of *The Danbury News-Times* or the public library and cannot be removed. By special arrangement, they may likely be inspected on the spot, but I should feel like a bee in a glass hive trying to read or write by special arrangement on somebody else's premises.

It's not that I'm temperamental.

For one thing, I am so slow and inexpert I'd be forever. I am also sensitive about this matter of being a writer. To myself, it appears presumptuous of me. What must it seem to others?

I confided my diffidence to a friend who instinctively said the right thing.

James Montgomery Bailey

"Pooh! It's nothing to be ashamed of. Why, in this vicinity writers are a dime a dozen. Anyway, you don't have to be a writer just to write."

"I'd love to believe that," I said.

Thus emboldened, I should eventually have gone to the library and might be sitting there yet had not manna arrived from heaven to relieve my predicament.

Raven #1 contributed a nearly complete set of premium lists that have been preserved in his family for four generations. These contain accounts of the Fair's yearly expansion and progress.

Raven #2 proffered a voluminous scrapbook of old newspaper clippings obviously compiled by a former fair official. "There," he said, "don't ask me how I came by it. Keep it as long as you need it, but give it back." I seized upon it hungrily. In the silence of my own familiar haunts and in no more critical presence than the cat's, I read newspaper reporting and commentary on fairs of sixty and seventy years ago.

The process marched by slowly at first since I have what educators refer to as "a short attention span." These souvenirs, however, together with the *History of Danbury* proved my salvation.

2

MORE HISTORY STUFF

According to a census taken in 1860, Danbury's population was 8,234 and growing at a rate of about 500 people a year.

Prior to 1850, it must be presumed, Danburians did not get around much—such transportation as they had being provided by the horse, the bicycle and shank's mare.[13] But in 1852, the Danbury-Norwalk railroad was completed and published a timetable offering two trains a day in each direction. Those from Danbury connected with trains for New York and "for the East." "The East," I suppose, meant Bridgeport, New Haven, Hartford, Boston and other way stations.

Travelers were able, for the first time, to get out of town without making the six-mile trip by stage to Hawleyville, where the Housatonic Road from Bridgeport to New Milford had a junction.

13 An expression that refers to walking or making use of one's own legs.

This Housatonic Road, as described by J. M. Bailey, left much to be desired beyond accessibility.

"The road was in a crude state, of course. The rail used was an iron strap nailed to a timber. Occasionally it would happen that at a joint, an end of one of the rails would become loose and accidents of a serious nature frequently arose from this cause. The point of the rail would be pushed through the floor of the car, bringing death or serious disfigurement to the passengers in the way."

"Of course," the man says, and "frequently!" No wonder Connecticut became the cradle of life insurance.

Danbury followed the lead of larger communities with the introduction of illuminating gas in 1857. It was such a convenience that three years later twelve street lamps were ordered by the borough.

The third move in the right direction was the completion in 1860 of a municipally-owned water works at a cost of $37,500. "Nine miles of ten-inch water main," says Mr. Bailey, were laid through all the principal streets.

In spite of the new railroad, street lights and water works, life here until the turn of the century, as indeed all over America, remained geared to the speed and strength of the horse, not only for purposes of local transportation, but for the production of all crops. The family horse was as common, even among town dwellers, as the family car is today. In such affectionate esteem was he held, that a horse thief was regarded as the meanest and lowest kind of thief, much worse than the automobile stealer of our generation. His crime more nearly approached the wickedness of kidnapping.

Prior to 1879, there were no telephones at all here, not to mention long distance service or state police. Danbury had an active Anti-Horse Thief Association that co-operated with the Jefferson Valley Association and others around the area in tracking down these wretches. The members, called "riders," were called out when a report of horse stealing came to the local authorities. They frequently traced a stolen horse up hill, down dale and through swamps to recover him and arrest his rider.

In an economy where horse-flesh was so important, the knowledge and methods of breeders who had learned by experimentation how to produce and perpetuate desirable characteristics were widely put to use by owners who sought to improve the quality of horses for a specific purpose. There grew up across the nation a chain of great stock farms liberally capitalized and using modern methods. Smaller breeding enterprises were undertaken by many well-to-do farmers or wealthy sportsmen whose pride in the ownership of a fast horse for fashionable driving or racing was increased by the further satisfaction of raising his own colts on his own land.

Within easy access to Danbury, there were several good-sized stock farms. The Brill Stock Farm produced Percherons, those stylish draft horses of Norman French descent. Brookside Farm in Norwalk, owned by M. H. Parsons, specialized in driving horses. The great majority of the breeding establishments, however, were devoted to trotters. Of these, the most prominent in Danbury was the Ridgewood Stock Farm, owned by Samuel H. Rundle and George C. White on the land now occupied by the Ridgewood Country Club. Dotted about the countryside were other farms where the harness horse reigned supreme, namely, those of G. H. Beard of Bethel, T. S. Hoyt of Ridgefield, C. J. Peck and Oliver Northrup of Newtown, Jesse James of Hawleyville, and J. B. Merwin of New Milford.

Before the Danbury fairgrounds existed as such, there was a track for trotting on a portion of the level land that it now includes. There in November of 1860 the great Flora Temple raced against the Widow Machree, winning in three heats, as chronicled by Hiram Woodruff in his *Trotting Horse of America.*

Horse racing, formerly in disrepute in this country as a pursuit of unsavory characters, had gained a surge in popularity after the Civil War. The opening of Central Park in New York City not only awakened the world of fashion to the advantages of vehicular display and made riding by ladies, sidesaddle of course, socially acceptable, but it brought about the reappearance of trotting as a gentlemen's diversion.[14]

While society leaders were vying with each other for possession of the finest driving horses to draw their barouches and victorias, wealthy men of affairs were bidding up the prices on fast trotters, whose speed and stamina they delighted to compare.

A far-famed rivalry sprang up between Commodore Vanderbilt and Robert Bonner, publisher of the New York Ledger and owner of the celebrated trotting horse, Dexter, whose time for the mile was 2:17¼, which was then a world record.

Both were ardent horsemen with wealth and leisure to spare for the cultivation of such an expensive hobby, the most conspicuous difference between them lying in their respective attitudes toward the placing of bets. Commodore Vanderbilt lustily delighted in backing his trotters to win the well-publicized contests known as "brushes" that took place in Harlem Lane, as St. Nicholas Avenue was then called. However, Mr. Bonner was a conscientious objector to gambling in any form, even refusing to race his horses because of his scruples in that connection.

In 1866, Leonard Jerome, the maternal grandfather of Winston Churchill, built his elaborate race track, Jerome Park, just north of New York City near Fordham, as headquarters for his Jockey Club. This was patterned after the ultra-exclusive jockey clubs of England and France. The park's inaugural was attended by the cream of New York society, including not only the new smart set, but members of the very old conservative families. Among those present was General Grant, the national hero, who three years later, in 1869, was to become our 18th President.

14 Lloyd Morris, *Incredible New York*

The influence of such prominent financial and social leaders was felt far beyond the limits of New York and Westchester. Gentlemen's driving clubs were organized in almost every community and hardly a village was without its half-mile track. All over New England and the Middle Atlantic States the trotting park began to exchange its shady status for a position in the full sunshine of respectability, as if a sport is to be judged by the company it keeps. Only among church people was this light somewhat filtered by distrust as they saw their fellow citizens congregating in small groups here and there along the track's side "a' putting their hands in their pockets and a' taking them out."

It was on the crest of this change in sentiment that the Danbury Pleasure Park originated.

From the *History of Danbury* comes the following paragraph:

> *In the spring of 1869 Messrs. S. H. Rundle and Jacob Merritt bought the grounds now owned by the Danbury Farmers' and Manufacturers' Society. Soon afterward they associated with themselves George C. White, Benjamin Lynes, and George W. Cowperthwaite and under the name of Danbury Pleasure Park laid out the present grounds and track. These were two classes, one for a purse of $175, open to all horses, in which the fastest heat was 2.40, and one for a purse of $150 for 2.50 horses.*

By present day standards the purses were small and the horses were slow, but this was still rather early in the annals of American trotting. Purses have increased in these eighty years and, thanks to the men who have bred, trained and cherished them, horses go faster too as if to be worthy of the increment.

Public interest in trotting being what it was and the Fairfield County Agricultural Society having definitely committed its fairs to Norwalk by the purchase of grounds there made it was natural for Danburians, who have small taste for being left out of an act, to think to themselves that a fairgrounds goes with a race track as surely as hay is for horses. Here was an already established trotting track in the midst of a level 30-acre tract. An annual town fair would advertise and bring patronage to Danbury, the town fathers decided.

Messrs. Rundle, Merritt, White, Lynes and Cowperthwaite were sounded out and found agreeable to a proposal that, if carried out capably, might become mutually profitable.

Accordingly on August 5, 1869, there appeared a short notice in *The Danbury Times* that read, "A meeting of citizens was held on Saturday afternoon last to talk up the subject of a town fair, Lyman Keeler in the chair and J. M. Ives, Secretary."

The meeting resulted in the founding of the Danbury Farmer's and Manufacturer's Society. By September 4, arrangements had been made with the Pleasure Park Association for the use of their grounds by which each group was to

assume one half the risk and to share equally in the profit or loss of the Fair.

1869 was the fourth year of the period of reconstruction after the Civil War, and so not particularly auspicious for a new enterprise. True, the month of May had seen the driving of the "Golden Spike" to mark the completion of the Union Pacific Railroad, which was later to do so much toward unifying the nation. But September 24 of that year has come down in history as Black Friday, the day when Jay Gould and Jim Fisk, Jr., tried to corner all the gold in the banks of New York City. When the government broke the corner by placing $4 million in gold on the market, falling prices and financial panic followed.

Perhaps sixty miles from Wall Street was too great a distance for the repercussions of Black Friday to be felt immediately in Danbury. At any rate, the first Danbury Fair had its opening twelve days later on Wednesday, October 6. Tuesday, according to *The Danbury News*, was "devoted to the reception and arrangement of exhibits. The public days were Wednesday, Thursday and Friday."

On October 11, the Treasurer's report, listed gate receipts at $1,950.73 and income from other sources S423.00—a total of $2,373.73. It was thought that after the bills were paid there would be a small balance in the treasury.

In 1870, encouraged by this modest success, the Danbury Farmers' and Manufacturers' Society and the Pleasure Park Association united to form a joint stock corporation called the Danbury Farmers' and Manufacturers' Company. The following year, its title was changed at a stockholders meeting to "The Danbury Agricultural Society" and so it continued until November 19, 1916, when its corporate name was changed again—this time to "The Danbury Fair."

The first Fair opened in a single tent borrowed from the Barnum and Bailey circus, but over the years as capital was available from profits buildings were added until "in 1897, there were for exhibition purposes the main building, 105 feet long and 90 feet wide with two wings, one for an art gallery and one for display of machinery, a poultry building and a bench-show building. There were also a wooden grandstand, which would seat 5,000 people, a caretaker's house, sixteen stables for trotters, and a secretary's building where the business of the Fair was carried on."[15]

15 J.M. Bailey, *History of Danbury*

Three additional structures for housing displays were formed by wooden walls, over which canvas tops were stretched to provide shelter.

From its inception, the Fair was well and profitably conducted by its founders, their families and their friends. Blocks of Fair stock were originally held by a tight little group who influenced about 300 acquaintances to buy a few shares each thus cannily providing working capital and assuring local interest. Salaries of officials varied from year-to-year depending upon profits, and the management tried to keep a small surplus in the treasury with which to meet emergencies.

A postcard depicting the Administration Building, Big Top and Grandstand, along with a hospital tent, in the late 1800s. *Danbury Museum and Historical Society*

3

FOUR-LEGGED BEGINNINGS

The great contribution of the agricultural fair to this country's progress has been the constant betterment of the breed by fostering competition. That mainspring of American character functions best when there are specific rewards for excellence, but a first prize, whatever its cash value, is primarily a token of fulfillment. The fair is the proving-ground where a breeder is able not only to parade his own accomplishments, but to compare them, and so satisfy his natural curiosity into the achievements of his peers. Win or lose, he comes home a wiser man.

Not all losers, however, are temperamentally inclined toward the long view, and some habitual winners react unreasonably to a challenge. Clashes between such intolerant personalities commonly lead to the choosing of sides and a general embroilment, which displeases nobody.

Long before the Bonner-Vanderbilt affair, song and story had established public interest in a good feud as a certainty. In modern times, "publicity by grudge" is known as a first-rate device for boosting attendance or making sales. Harvard and Yale are familiar with this technique, as are Gimbel's and Macy's.

Ready to the hands of whatever press agents were called in those days was a bona fide rivalry between the owners of the two leading stock farms in this area. During the 1890s, contention between Sam Rundle's stable and that of William Beckerle was the spice of Fair Week, which was enriched and enlivened to no end by that stimulating ingredient.

Sam Rundle was that same S.H. Rundle who, with Jacob Merritt, purchased the land for the original Danbury Pleasure Park. He was a local hat manufacturer and banker who had achieved notable success in both of these capacities, but whose pursuit of happiness followed the flying heels of his high-bred trotting horses.

His prize stallion, Quartermaster, was the best known among the trotting sires of this area, having acquired prestige not only for his beauty and highly-finished performance, but for having won at the New York horse show.

Of him in his prime, *The American Horse Breeder* had this to say:

> *Quartermaster (2.21 ½) ranks as decidedly the leading son of Alcyone (2.27), which is regarded by many as one of the very best of the sons of George Wilkes. Quartermaster is also fashionably bred on the maternal side, his dam being Qui Vive (dam of Guardsmen, 2.23 ½), by Sentinel (2.29 3/4) a successful sire and full brother to the famous Volunteer; second dam, the great brood mare Missie (dam of King Wilkes, 2.22) by Brignoll (2.29 3/4).*
>
> *He is a handsome horse and not only makes a fine appearance in the show ring, but so uniformly are his qualities transmitted that it is easy to select from his offspring any requisite number for show purposes.*

Trotters at speed, circa 1934. Note the low-wheeled, rubber-tired sulky—an enhancement adopted from bicycle racing. *Frank Baisley*

Ridgewood Stables housed a score of Quartermaster's sons and daughters, while for miles around his progeny was well and favorably known.

William Beckerle was a rival hat manufacturer who owned the rival stock farm. His career had been spectacular.

In 1866, at the age of 20, he arrived from Germany and first went to work as a farmhand for Lewis Elwell on Clapboard Ridge. It didn't take him long to see that hatting was the thing in Danbury, and he set out to learn the business by getting a job with the Tweedy Manufacturing Company. He learned so fast that within ten years he had established his own firm of Beckerle & Co. and soon afterward bought for himself the Elwell place where he had first found employment. On its 280 acres, he built a fine home complete with stables and training track for his horses. He called it "Hilltop Farm."

"Three generations from shirt-sleeves to silk hats" went an old saying, but Mr. Beckerle made it in one, his own. Later in life, he sustained severe losses, when first his main factory and afterward his stables were wiped out by fire, but his influence on the industrial life of the community was long-lasting. The Beckerle Hose Company is a present reminder to us Danburians of his interest in civic affairs.

Sharp, sometimes bitter, was the competition between Ridgewood and Hilltop.

The citizens of Danbury were pretty evenly divided in their allegiance. Bill Beckerle was a parvenu[16] in this Revolutionary settlement, but his integrity and known generosity, especially in the matter of loans to families of German descent, had made him many friends. These supporters were not, as a rule, associates of Mr. Rundle, who represented the old guard. Thus it came about that the big wooden grandstand bristled with partisanship as season-after-season Mr. Beckerle matched his horses against those of the Ridgewood Stock Farm.

Hilltop had Sablenut, Sablehurst, Sablewood, Sableur and Beverly, who were all sons of Sable Wilkes, he by Guy Wilkes, by George Wilkes, whose sire was Hambletonian. Through Lady Bunker, dam of Guy Wilkes, these colts were also descended from Mambrino Chief.

Furthermore, Hilltop possessed the black stallion, Onwardo, "son of Onward (2.25 ¼), dam Lady Thorne (2.25) by Darlbay, son of Mambrino Patchen."[17]

Onward was head of the studs at the Patchen Wilkes Farm in Kentucky. His dam, Lady Thorne by Mambrino Chief, was "next to Dexter, the best trotting horse of her time."[18]

Thus the two great lines of Hambletonian and Mambrino Chief were joined in the ancestry of Hilltop horses.

Fair Week after Fair Week, it was Quartermaster and Onwardo in the show ring, Ridgewood versus Hilltop on the track. Each in turn had its innings.

The two newspapers did their best to chronicle the warfare impartially and their circulation rose as horsemen and hatters, who couldn't tell a trotter from a pacer, displayed equal interest in the particulars of each day's skirmish.

---

16 A person of obscure origin who has gained wealth, influence, or celebrity.

17 According to *American Horse Breeder.*

18 From Dwight Akers in *Drivers Up: The Story of American Harness Racing.*

In October 1894, *The Danbury News* published accounts of events at the track, which furnished a fair sample of the nip-and-tuck contests:

"Sablenut's Performance"

> *On Friday of Fair Week, an exhibition mile was trotted by Sablenut, the two-year old stallion of the Hilltop Farm, which was no doubt the best mile ever trotted on this track, and the fastest mile ever trotted over any half-mile track in the country, by any two-year old.*
>
> *About 4 o'clock he came out and, after being driven around the track once, he started with a running horse beside him and trotted the entire distance without as much as a skip, the first quarter being made in 36 ¾, the half in 1.12 ½, the three-quarters in 1.481 and the mile in 2.23 ½, the last quarter being made in 35 seconds, which is a 2:20 clip. To those who witnessed this exhibition it was a rare treat, as the colt went so even and true and without any great effort or distress.*
>
> *Sablenut is one of the best bred colts in New England, being by Sable Wilkes, he by Guy Wilkes, which is one of the greatest Wilkes strains there is. His dam is Auntie, by Dawn, and his second dam Netteo, by Anteeo, by Electioneer, a combination of Wilkes, Nutwood and Electioneer, being the three greatest families of the day.*[19]

On Saturday, the next day after this exploit of Sablenut's, there was another story to tell about the track's record. This time a son of Quartermaster received top billing:

"A Sensational Mile"

> *The last performance of the record-breaking fair was a record-breaking mile by Quartermarch. On Saturday, after trotting one mile in 2.41 ½ and another in 2.53 ¾, which was fast enough to keep in front of the other two starters in the 2.50 race, Quartermarch was sent against the track's record. He duplicated the old George Wilkes action as to style and went several points better as to speed, passing the quarter in 35 ¼ and the half in 1:10 ½. The next half was at an increased rate. The watchers noted 1:45 at the three-quarters and 2:19 ½ at the finish. How many seconds he can cut off this performance no one knows.*[20]

And since this was the last performance of the season, nobody knew how much better Sablenut could do, either. Quartermarch had had the last word.

---

19 *The Danbury News*, October 7, 1894

20 *The Danbury News*, October 8, 1894

So, with many an up and down, the feud of the stock farms continued to the delight of the grandstand customers and the benefit of the Fair. Where local interest was aroused, good fields were assured.

Within the fraternity of horsemen, Danbury's reputation was good. Purses were felt to be liberal and the level of trotting was high.

Since "the spring meetings eliminated fifty percent of early starters which failed to show speed, and twenty-five percent more were returned to the stable before the fall trots,"[21] only the fittest horses were, therefore, left to compete in the Danbury presentation, which wound up the season of fairs in New England.

This was one of the prime considerations that kept the Fair to its original first week of October dates, as year after year track records were whittled away by seconds and quarters of seconds. Between heats, it was customary to entertain the waiting spectators by various performances on the track or on a wooden platform just inside it.

Peg and umbrella races were sure-fire since the contestants were generally well-known.

In the peg race, there were usually about half a dozen entrants who raced their horses once around the track, where they stopped before the grandstand and unharnessed, hanging the harness on a peg or over the fence. They then re-harnessed and made another round of the track. The first to complete the routine was declared the winner. The fun lay in the mad scramble in front of the stand.

The umbrella race was much the same except that it was a running race in which each contestant had to dismount, light a cigar, raise an umbrella and remount his horse. This feat called for some practice since the cigar had to remain lighted and horses dislike umbrellas. What an ideal entrant Groucho Marx would make!

There was even a race between Albatross, the guideless pacer, and Samuel Wheeler, a gentleman of this city, on a bicycle. How well-named Samuel was!

"It was a pretty sight as they came down the back stretch, the horse pacing as steadily and making as great an effort to win as if being urged by a driver, and it was an exciting finish as they came to the home stretch neck and neck, the horse's head even with the wheel of the bicycle."[22] The race was won by Albatross, whose time for the half-mile was 1:06.

In 1895, the bay and white trick mare, Mazeppa, answered questions put to her by the audience by nodding or shaking her head. She also counted by striking the ground with her foot. When asked how many days in the week she preferred to work, she shook her head vigorously.

To the sorrow of thousands who had admired her, she and her attendant, George

21 *The Danbury News*, October 5, 1898

22 Ibid.

Longhill, were killed in a train wreck near Waterbury on the way to Manchester, New Hampshire, the day following the close of her Danbury appearance.

The most acceptable offerings were usually the acts involving animals, Adam Forepaugh's educated horses and Hagenback's lions being stellar attractions in various years. Other attractions were tumbling, juggling and acrobatic acts.

For several seasons, Rube Shields, a trick cyclist who had won fame by "his daring and perilous ride down the West steps of the capitol at Washington, D.C,"[23] gave exhibitions of fancy bicycle riding and entertained the crowds by riding down ladders on his wheel.

In 1896, the first horseless wagon to be seen in Danbury was exhibited on the track, doing the mile in 3 minutes, 42 seconds.

"Something for everybody" was the avowed ambition of the management, but however calculating these dry New Englanders were in the conduct of their hobby, they were motivated by a real love for horses and cattle, a fervor that infected their neighbors and other livestock owners for miles around. Race meets and the display of blooded stock were a culmination of long striving, and a ribbon—blue, red, yellow or white—was a much coveted badge of honor.

During the 1880s and 1890s, the Danbury Fair became outstanding in New England for quality of stock and judging.

Champion bull, 1919. *Frank Baisley*

23 *The Danbury News*, October 8, 1897

4

UPS AND DOWNS

From 1870 to the close of the century, Yankee ingenuity was being ridden full-tilt toward the development of those conveniences that we now regard as necessities. As inventions were perfected, they were displayed and demonstrated wherever large groups of people congregated. Indeed, after 1869, all the major inventions and most of the minor ones were represented in accounts of the successive Danbury Fairs.

For example, the first practical telephone was created by Bell in 1876. In 1879, it was introduced into Danbury, which would appear to be rapid exploitation of a facility, even for these days.

By 1895, the installation of a complete telephone system on the fairgrounds attracted much attention and comment.

"There was in operation," remarked the *News*, "a central station, a switchboard and thirteen automatic pay stations. Besides being a great convenience, the system will be a very interesting exhibit."

Edison invented the first incandescent electric light bulb in 1879. Danbury waited longer for electricity than it had for the telephone, but by 1887 the new lighting was turned on. At the fairgrounds, the Administration Building was wired first, then others in order of their importance. By 1896, the management was planning to light the midway for an evening carnival "making it a miniature Coney Island."

In 1894, Edison's latest invention was described by a Fair reporter as follows:

By this device a series of minute photographs of living objects in motion were shown. The photographs, presented on an endless band of film which passes under the field of a powerful magnifying glass, were flashed before the eye at the almost incredible speed of 46 a second. So rapid is the change that the eye cannot detect it and to the observer these successions of separate and distinct photographs present the illusion of living, breathing forms.[24]

Another item read:

Edison's kinetoscope is a favorite with most people and the two shown in the tent are well-patronized. One shows a butterfly dance and the other a rooster fight.[25]

The "movies," which were to dominate the amusement industry during the first half of the 20th century, were getting under way.

The Fair was already a magnet for farmers and breeders of livestock. Now its drawing power attracted the average citizen as well, eager to

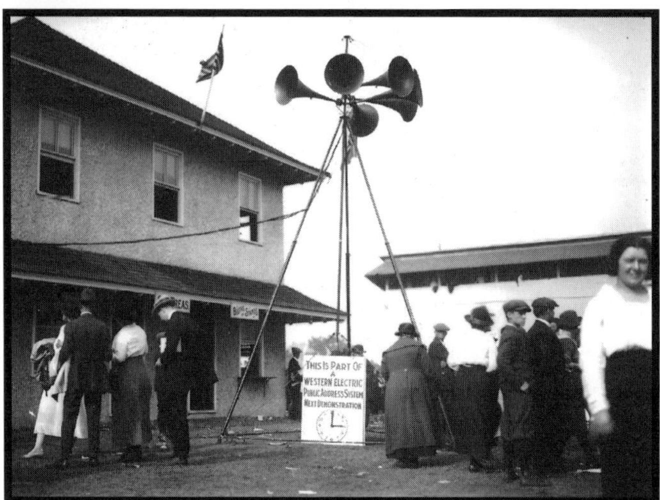

New marvels of science and invention on display at the Danbury Fair.
Top: An early flying machine from the 1930s.
Bottom: The sign reads, "This is Part of a Western Electric Public Announcement System. Next Demonstration 3:00."
Frank Baisley

24 *The Danbury News*, October 1894
25 Ibid.

glimpse the new marvels of science and invention. In this way, the rapid progress and general liveliness of the period contributed to the Fair's growth.

Grow, it did, from the very beginning.

In a preface to the premium list[26] of the third Danbury Fair, shares of stock in the Danbury Agricultural Society were first offered for sale at $25 each, a payment that entitled the holder to an interest in the company, a season ticket to the Fair and the use of the track during the year when not otherwise occupied.

The offer was repeated in the lists of the fourth and fifth years, but was omitted from that of the sixth and all subsequent publications, leading to the conclusion that in three years' time the stock issue had been fully subscribed.

In place of advertising Fair stock, the publication of the sixth year substituted the following paragraph:

> *The success of the Society since its first formation has been most remarkable. Organized at that time by the efforts of a few public-spirited gentlemen, to supply what was deemed a public need in this vicinity, without capital or funds, it is now, in the sixth year of its existence, the owner of its admirably arranged fairgrounds, containing one of the best half-mile tracks in the country, is entirely free from debt, and receives annually for admission fees and disburses for premiums a greater sum than any other similar organization in the STATE.*[27]

In 1875, the seventh year, we find a tabulation of the "receipts of similar organizations in the State as Reported to the Board of Agriculture (not including State appropriations)."

Watertown	$2,780.28	Meriden	$2,797.03
Wallingford	$2,281.18	Norwalk	$2,221.02
Tolland	$1,878.10	Ridgefield	$1,907.16
New London	$2,700.05	DANBURY	$4,919.14

As evidenced by its receipts, this descendant of the Fairfield County Agricultural Society had in six years attained twice the popularity of the Norwalk Fair, a superiority that continued to increase until 1888 when the County Fair Association gave up the ghost and sold its grounds in Norwalk.

The *Official Report of the State Board of Agriculture, on The Danbury Fair of 1887* comments:

> *It is stated as a fact that more Norwalk people attended the Danbury Fair than attended their own County Fair in Norwalk, and the Fairfield County Fair at Norwalk was no doubt as good a county fair as was held in the State.*

26 A program booklet that outlines the various categories, requirements and awards for a contest.

27 *Danbury Fair Premium List*, 1874

The Report concludes with these sage observations:

> *There are three very essential factors necessary to running a successful fair! 1st, Plenty of money to furnish suitable accommodation for exhibition; 2nd, Good practical men who can appropriate all the time necessary to work it up; and 3rd, Good weather. With these, many more agricultural fairs can be made both large and successful, but to outdo or equal the Danbury Fair, it is doubtful if it can now be done in this State for many years to come.*

As the years passed, there were attempts to establish fairs in much larger areas of population about Connecticut, but little or nothing came of them to detract from the drawing attraction of Danbury. A State Fair that was opened at Hartford's Charter Oak Park received from the legislature a yearly appropriation of $3,000, but it failed to survive. It was later revived in Meriden and there petered out.

The State of Connecticut recognized early the contribution of the agricultural fair. Besides the $3,000 to Hartford, it paid varying amounts in 1878, averaging $124 each, to other fairs throughout the State.

The directors of the Danbury Fair were of two minds concerning these handouts. State money for premiums was certainly desirable, but they suspected that its acceptance might lead to complications.

Consequently, there were years when they accepted the State's bounty, and other years when they failed to apply for the appropriation.

"Too much red tape," was the answer to explain the unusual attitude represented by the latter course, but that was only a camouflage to mask a wary unwillingness to leave the door of the Fair far enough ajar to risk the admittance of a political foot.

The number of classes for which cash premiums were given increased from twenty-six to fifty-five over the first twenty years of the Fair's existence, and the value of the cash paid rose from $1,500, "a sum which seems large, but which we are persuaded the success of the exhibition will justify,"[28] to $4,000 in 1889.

By 1888, it was generally conceded that the Danbury Fair was the banner fair in the State. Newspaper commentary was almost all favorable. The waggish editor of the *Norwalk Hour* remarked:

> *The Danbury Fair, which promises to be larger and better than ever before, if such a thing is possible, opens Tuesday of next week. We have heard of just one Norwalk man who will not be in attendance. He is old and lame and blind, and also confined to his bed by a stroke of paralysis.*

The *South Norwalk Republican* had heard that "there are twenty-five persons in Danbury that will not attend the Danbury Fair. They are all in jail however."

---

28 *Danbury Fair Premium List*, 1889

Except for the minor misfortunes of a few rainy days, the Fair ran along without serious setback for almost thirty years, but it is not always May.

On October 21, 1897, *The Danbury News* reported the first calamity to befall the Agricultural Society.

> *Fire swept over the fairgrounds at midnight last night and laid in ashes all the principal exhibition buildings and by 1:30 o'clock this morning there remained only an acre or two of blazing debris to mark the site of the old familiar landmarks. The fire started apparently in the corner of DeKlyn's Candy Stand.*

James Nolan, the caretaker and watchman who lived in the farmhouse to the north of the Big Top, was awakened at midnight by an explosion. He aroused his family, jumped into his pants and rushed outside to find flames bursting through the ventilators on the roof of the main building.

There was no firefighting equipment on the grounds and, in a few minutes, the whole enclosure was seething with flames. The northeast wind, blowing hard, swept the fire fast and furiously through the long passage ways.

Nolan led the horses out of the nearby barn and, with the help of a dozen people who had seen the fire from the road, saved what he could from the secretary's office, which was the last to catch fire.

From the main building, the wind carried the fire first to the poultry building and next to the bench show building. Agricultural hall was then consumed and after that the smaller structures to the West.

The *News* reporter wrote: "From the hill at the top of Park Avenue the road was light as day and the scene one of weird grandeur. At half-past twelve it was possible to read a newspaper in any part of the city west of Main Street."

Fair officials suspected that the fire was of incendiary origin since they felt sure there was nothing in the building that could have caused an explosion.

Lost with the buildings were the huge tents stored inside them as well as expensive showcases, turnstiles, flags and other ornaments.

The wooden grandstand and Nolan's house were spared because of the direction of the wind. The cattle buildings on the other hand, constantly showered by sparks and burning embers, were only saved by the constant watchfulness of neighbors who guarded them with brooms and buckets of water.

A rough estimate placed the loss at $15,000 on the buildings with a further loss of $5,000 on their contents. However, the buildings were only insured for $11,000.

Thankfully, 1897 had set a record with a profit of $6,000 in excess of any previous year. This money together with the insurance was available for the replacement of the burned structures.

S. H. Rundle had at this time been president of the Society since 1892, the same year he was chosen president of the Danbury National Bank. It was his decision, unanimously seconded by his directors, to rebuild.

W. H. Osborne was retained to draw up plans and specifications and the Society advertised for bids on material and labor.

In the spring, work started early and the Danbury Fair began to rise again from its ashes. Its new buildings, more spacious and convenient than before, were again ready by the following October to shelter the products of field and farm.

The Fair's recovery from this disaster was complete and subsequent triumphs were recorded, for which most news writers gave credit to the management.

> *Agricultural Fairs are held in one sixth of the towns of the State, but very few of these exhibitors aspire to the dignity of expositions, and indeed the Great Danbury Fair is the only annual event in New England which attracts any considerable attention from the outside world. During the last score of years the managers have conducted this enterprise with so much wisdom that the Danbury Fair is now acknowledged to be the most complete exposition of its kind ever held under canvas in the United States.*[29]

To the management indeed praise was due. Its success lay in teamwork—of a particular sort.

The history of the failures of fairs in New England is the story of disagreements among their directors and the quarrels that ensued.

The officers of the Danbury Fair appeared to work hand-in-glove with each other. There are records to show that differences arose from time to time, but the fabric of the Fair, in spite of a few rips, was never completely torn apart by dissension. At the top, there was a strong guiding hand and intelligence popularly acknowledged to belong to Sam Rundle. The Fair had sprung from his purchase, with his partners, of some land for a trotting park and it was no part of his plan to let others dictate general policy. The inner circle of his close friends and blood relations held key positions as officers and served on the board of

Father and son, Samuel and G. Mortimer Rundle. *Frank Baisley*

29 *New Haven Palladium*

directors. Up-and-coming young men with a flair for management he made into superintendents of departments where their self-respect was gratified and their energies were fully put to use.

Now the iron hand in the velvet glove is fine, unless the glove slips. Sam Rundle's fellow townsmen occasionally saw or thought they caught glimpses of dictatorship. Nothing is more infuriating to the free spirit of a proper Danburian than to feel he is not privy to what goes on.

At the annual meeting of stockholders in 1898, there was the making of one of those controversies that have split many a fair association wide open. The local papers made no mention of internal strife, but the *Waterbury Herald* headlined it:

A FAMILY AFFAIR.

> *For more than a year there has been a sentiment of unrest and a growing feeling on the part of many of the stockholders that it was time for the Fair Association to pass out of the hands of the one family which now controls it and be run for the interest of the many rather than the few.*
>
> *They claim that Samuel H. Rundle, the president of the association, is running matters to suit himself and that he has gotten all the principal offices into the hands of members of his own family.*
>
> *Up to five or six years ago, Rundle owned but five shares of the stock, but when the estate of Benjamin Lynes was settled up he purchased seventy shares of Fair Association stock. With the seventy-five he now holds, there goes considerable prestige.*
>
> *One point which particularly irks the minority is that whenever Rundle's horses trot, he is, they say, always in the judges' stand directly behind the judges during the race. It has been suggested that the name of the organization be changed from the Danbury Fair Association to Rundle's Stock Farm Exhibition.*

The dissatisfied members pointed to the list of officers for 1897 to maintain their argument. Samuel H. Rundle was president, his son, G. Mortimer Rundle, was secretary and his son-in-law, John W. Bacon, was treasurer. The executive committee consisted of S. H, Rundle, J. W. Bacon and G. M. Rundle. The board of directors, consisting of ten men, including the same names and that of James W. Porter of Danbury who worked with the Rundle family. John W. Bacon was superintendent of building exhibits; Mrs. John W. Bacon, of the ladies industrial department; J. W. Porter, of tent exhibitions; G. M. Rundle was one of the two men at the head of the horse department and also of the cattle department; S. H. Rundle was one of the superintendents of the booths, stands and police departments.

The piece concludes:

> *At the annual meeting last Saturday night, an attempt was made to swing enough votes to break up this combination but it was unsuccessful. When the final vote was announced the majority for the old board was 298, but the total number of ballots cast exceeded by 52 the number of original shares of Danbury Fair stock.*
>
> *The minority are at a loss to know where the extra 52 votes came from.*

Issues among the stockholders weren't the only problems the Fair had to suffer through, though. In 1916, the 48th annual premium list carried a special notice printed in red ink and inserted inside the cover as a first page.

> *The main exhibition building and the offices on the Danbury fairgrounds were destroyed by fire following a stroke of lightning on Sunday July 9. It has been found to be impossible to rebuild anything but the offices in time for the Fair of this year. Therefore, with exceeding regret, the officials are compelled to announce that, as they have no place to house them, it will be necessary to eliminate the Fine Arts, Ladies' Industrial and Merchants' Display departments. The Main Tent and every other department of the*

Aftermath of the third fire at the Fair, 1922. The Grandstand was a total loss. *Frank Baisley*

> *Fair will be ready at the appointed time, a new tent having been ordered and the necessary connections made for it.*

By the next year, the main building was completed. However, the Fine Arts display was never revived. The Ladies' Industrial went into a decline from which recovery was long deferred. It wasn't until 1935 that needlework and articles of home manufacture were again exhibited.

A third bad loss by fire took place six years later in 1922 on Saturday, the last day of the Fair.

A certain William Britton had a lunch counter under the grandstand where he cooked on a woodstove with a smoke-pipe going up through the wooden framework of the stand.

I hate to think of what Lieutenant Sullivan of our State Fire Marshal's office would say to that makeshift arrangement had he been around at that time. Lt. Sullivan's professional interest is arson and he starts coming around to the fair-grounds early in September every year with a microscope in one hand and a blowtorch in the other. Every section of canvas and every yard of bunting must be fireproofed to withstand the scrutiny of his eye and the fire of his torch. In 1944, the authorities of the State of Connecticut had become understandably upset over the Barnum and Bailey circus fire in Hartford, a disaster in which 168 lives were lost. Inspections such as Lt. Sullivan's were implemented to avoid a repeat.

Yet the 1922 fire was before such measures were in place, and the old wooden grandstand burned flat. Its replacement before another Fair-time was crucial.

The new fireproof grandstand was ready in time for the opening of the next Fair. Photo circa 1924. *Frank Baisley.*

The Danbury News on May 15, 1923, carried a reassuring item.

> *Work is rapidly nearing completion on the steel framework for the new concrete stand at the Fair Grounds.*

The new grandstand was well-built of fireproof construction—420 feet long by 60 feet in width. It cost over $100,000 in 1923. Today it would take more than twice that amount to replace it. It was planned to seat 6,000 spectators, but they have to be fairly well-acquainted, especially in hot weather.

Before World War II, Lucky Teter used to fill it Saturdays and Sundays with his thrill show, and it was filled to overflowing for midget auto races in 1946 and 1947. Trotting races never did quite so well. For these, it was cheaper and actually more fun to be a railbird.

Fortunately, the fire of 1897 took place just after, rather than during, Fair Week, so that there was no injury to man or beast.

Again there was insurance money, but the cost of enough fire insurance to afford substantial protection has always been prohibitive because of the distance from a firehouse and because the main water supply from Lake Kenosia is shut off during freezing weather. The several good wells on the property couldn't begin to furnish enough water for firefighting.

Third from left: George Nevius, secretary. Fourth from left: Governor Wilbur Cross. Second from right: G. Mortimer Rundle, president. *Frank Baisley*

The financial loss was greater this time, but still not disastrous. It never does any harm to have a bank president or two and as many directors as possible associated with an undertaking of this kind.

This is not, John wishes me to insert, that the resources of the banks were at the Fair's disposal, but that the officers who made the loans were familiar with the value of the property and its earning power!

The Rundles and their associates were in a position to know.

As Sam Rundle grew old, one of his greatest satisfactions of life was seeing most of his duties and interests capably undertaken by his son, G. Mortimer Rundle, who had already been mayor of Danbury as well as an officer in the Danbury National Bank and secretary of the Agricultural Society.

When the father died at the age of 95, the old order continued with few visible changes in the conduct of the Fair.

In his turn, Morty Rundle allied with himself those tried and trusted associates among whom no serious disagreements were likely to arise. The pattern was there for him to follow and he followed it when he became president with the selection of his own son-in-law, Stuart McLean, as vice president, and his old friend, George Nevius, as secretary.

These were, I suppose, the offices that really counted since the treasurer did not shape policy. All he had to do was handle the tickets, count the money, do the bookkeeping and be above suspicion. Personally, I think the treasurer's office is the heart of the Fair, and that an honest treasurer is to be valued more highly than rubies. Maybe Mr. Rundle thought so too. He enlisted some good ones.

The Rundle combination never was broken up, but its despotism was benevolent. Somewhat smugly conscious of work well done, the officials paid about the same attention to insinuation, suspicion and jealousy as Quartermaster slowly switching his tail in a field of clover paid to the buzzing of the horse flies.

In reality the interests of the minority were well served by the Rundle clan, who kept the Fair in existence despite fire, war and financial panic, regularly paying themselves small salaries or none in a year when profits were washed out by rain or when unusual expenses called for retrenchment. They worked hard and long for love of their hobby and the glory of the Danbury Fair. Ample evidence of their thrifty husbandry may be found today in the Fair's old books and records.

5

HOW THEY CAME

In 1890, 53,745 people attended the Fair "from near and far." Since the population of Danbury in that year was 19,386, and presumably some Danburians were too old, too young or too sick to make it, perhaps two-thirds of those present came from some distance.

That they did so was due in great measure to the railroad excursion.

By this time, Danbury was becoming a rather important link in the railway network of western Connecticut and southeastern New York.

The Housatonic Railroad, originally built from tidewater at Bridgeport north to New Milford, had been extended to Albany. To connect with the Housatonic, the Danbury-Norwalk Road had laid a six-mile branch from nearby Bethel to Hawleyville.

Hawleyville was also the junction point of another early line, proceeding in leisurely manner from Litchfield along the Shepaug Valley and named "The Shepaug."

By 1892, the New York, New Haven and Hartford, then known as the Consolidated Road, was in the process of absorbing these three roads and establishing itself solidly in the Danbury area.

The New York and New England Railroad, organized in 1846 to connect Boston with the Erie Railroad at Newburgh on the Hudson, was first put through to Waterbury. There, its building became stalled for so long that everybody forgot its proposed destination. Finally, in 1881, thirty-five years later, its rails reached Danbury and entered New York State at the Danbury-Brewster town line.

The village of Brewster, New York, ten miles from the center of Danbury, was already serviced from New York City by two divisions of the New York Central, the Putnam and the Harlem. There, with the arrival of the New York and New England, it became a junction point.

Passenger traffic used to be popular with the railroads. From points everywhere along their lines, the companies offered low-rate combination tickets that included transportation and admission to the Fair.

The service was generally excellent. Specials[30] were spotted on sidings[31] just outside the gates, while excursionists arriving on regularly scheduled trains transferred at the Danbury station to a single track shuttle train consisting of a few cars shackled between two small engines pointed in opposite directions.

As the regulars pulled in to Danbury, passengers kept their eyes peeled for the shuttle train. If the special was already on its siding, they swarmed to the platform and sprinted pell-mell for seats.

There was no waiting. Within ten minutes, the shuttle had disgorged its exuberant cargo and its crew was manning another locomotive for the return trip.

In 1894, the Fair was open to the public for five days starting Tuesday, the 2nd of October. Monday was a day of preparation, upon which exhibitors arranged their displays and concessionaires readied their stands.

Thursday used to be the biggest day for excursions crowds. "Out-of-town Day," it was called.

According to *The Danbury News* of Thursday, October 4, published before that day's figures were available, "The attendance of Tuesday and Wednesday was larger than on the corresponding days of any other year. Today the attendance is very large, but on account of the storms last night and this morning, it is feared there will be a falling off from the attendance of Thursday last year."

Under the heading "Travel on the Railways," the same paper ran the following items:

> *Travel over the New York, New Haven and Hartford was remarkably heavy today, and all trains were heavily loaded.*

---

30 Trains that weren't part of the regular transportation schedule.

31 Railroad spurs that took specials off the main tracks so that they didn't cause a delay or back-up for regularly scheduled trains.

The regular trains had additional cars and in some instances were filled to the platforms.

The special from Norwalk brought in 600 people.

Three hundred passengers came on the Falls Village special.

The Bridgeport special which arrived at half-past ten brought 500 persons to the Fair.

The Bridgeport special which arrived at half-past eleven had 200 passengers.

The New Haven train came in with 500 people on board.

Over the New England road, the travel was just as heavy, and the special from Waterbury had ten well-filled cars.

The Hartford train had about 600 persons.

The trains from the west were heavily loaded.

Specials were run from Brewster and from the Putnam and Harlem divisions of the New York Central.

Trains from Fishkill and a special which connected with the West Shore road brought a good many passengers.

The West Shore Road, designed primarily to connect metropolitan New York and New Jersey with Albany, meandered from its southern terminal at Weehawken along the west shore of the Hudson River from which it took its name.

That year, for the first time, it made its bid for a share of the Danbury Fair-ward travel and, undeterred by any considerations of distance or difficulties of connection, had handbills printed and distributed.

Danbury! Danbury! Beautiful Danbury!
Danbury Fair
October 1st to 6th, 1894
An Opportunity!
Special Excursion Thursday, October 4, 1894
Via
The West Shore Railroad
(Connecting at Newburgh for the New York
and New England Railway)
Don't fail to see the renowned Danbury Fair
A Mammoth, Matchless Exhibition

There followed a timetable showing the hour of departure and arrival from each stop north and south from Albany to Weehawken.

The excursion was actually more of a bargain than a casual reader might discern, including, if the passenger was a resident of New York City, two crossings of the Hudson between that point and Weehawken and a couple more from Newburgh to Fishkill.

Wherever he came from, the purchaser of this combination ticket must get off at Newburgh, be ferried to Fishkill, and board the New York and New England for Danbury, where his car would be shunted to the fairgrounds' siding.

Even if connections had been made according to plan, it was no junket for the aged and infirm. That connections were not so made, was suggested by two articulate ladies who made the trip:

> *October 14, 1894*
> *S. H. Rundle*
> *Pres. Danbury Fair*
>
> *Dear Sir:*
>
> *In the interests of the Danbury Fair and visitors from distant points, I address this letter to you.*
>
> *The West Shore Railroad advertised an excursion to Danbury Fair for October 4th, via West Shore to Newburgh, thence by ferry to Fishkill, and from there by N.Y. and N.E.R.R. to Danbury. The West Shore excursion train arrived at Newburgh too late to connect with ferryboat for the ten o'clock train to Danbury and, as there was no other train till 5:10 p.m. for Danbury, we were obliged to hire a man to row us across, who made $1.60 in ten minutes. We caught the train and got a seat in one of the cars in which no smoking was allowed (?). A man came in with a lighted cigar and, after taking a few puffs, kindly allowed his cigar to die a lingering death. The next passenger to come into the car, sat down opposite us, lit a cigar and went to work at it with vigor. As the conductor was too polite to inform his passengers of anything that might be disagreeable, we were obliged to inform the young man that the smoking car was the next one forward, and after swearing at us with his eyes, he took the hint and allowed his cigar to die that same lingering death.*
>
> *After spending five hours at one of the best fairs we ever attended, we boarded the train scheduled on the West Shore handbills to leave the fairgrounds at 6 p.m. half an hour early and secured good seats in the third car, supposing that the two forward cars would be sufficient to reserve for the drunks and smokers, but not so.*
>
> *After standing the atmosphere of liquor and tobacco for half an hour we left our good seats preferring to stand in the rear cars, where, to our dismay, smoking was going on openly. There were plenty of ladies and*

gentlemen who like ourselves were forced to stand this company of roughs and toughs for three mortal hours.

When we arrived at the West Shore station in Newburgh, we were informed that no arrangement had been made for taking Danbury passengers south, and that the next train left for our town at 4:40 a.m. October 5th, but after some talk the agent telegraphed for instructions and told us that the 9:16 p.m. express would leave us at our destination.

But our cup was not yet full. After the conductor had taken our tickets and asked us if we were for Stony Point, to make sure, he did not speak to us again until to inform us that we had already gone half a mile beyond our town, and that we couldn't stop till we got to the next station, as the whistle rope was out of order, and he had been fooling with it for some miles back but couldn't make it work. This entailed a long midnight walk.

In the interest of the public, the Danbury Fair and the Railroads, you should do your share toward abating such frauds as the West Shore excursion to Danbury Fair proved.

Yours resp'y,

Miss E. H. Moore
Miss R. E. Moore
Stony Point, Rockland County, N.Y. Box 283

The Misses Moore paid $1.60 each for their tickets and, according to the timetable, it should have taken exactly two hours to travel from Stony Point to Danbury.

It would seem that the organizers of the West Shore excursion had not thought the thing through.

In spite of the rain, Thursday's attendance that year was 14,176. Friday was Danbury Day, when stores and schools were closed and when anybody so inclined could shoot a cannon down Main Street at high noon without endangering a single soul. That year it was the biggest day with 20,200 present and paid for.

The local people came on foot and on horseback, by bicycle and by buckboard, in coach and in cart.

For sociability of design, the omnibus, open to the breezes, was the people's choice. Built along the lines of a police wagon with seats on the sides facing each other, it had a rear door with one step up and long, vertical handles of which to catch hold.

Every carryall for miles around saw service Fair Week in company with the hacks and the hansoms. From the railroad station down Main Street to West they all rolled, stopping to pick up fares.

"Only fifteen cents," yelled their drivers, "Right away over, only fifteen cents."

The words of this bally became the accepted term for the carriages themselves.

"I caught a right-away-over in front of the post office," Aunt Martha would relate, "and who was aboard, but the Finetooth girls from out Beaver Brook way. What a visit we had together!"

Fine, festive conveyances, these right-away-overs, but by 1895 they had had their day. In that year, the Danbury Horse Railway finished laying tracks to the fairgrounds and the new electric road with a 10¢ fare was put in operation.

After that happened, they gradually ceased to be.

The family horse earned enough oats Fair Week to keep him through Easter without lifting a hoof. The Agricultural Society made provision for him with posts and rails set up inside the track and around the outside of the grounds. A hitching charge of 25¢ was made for all teams admitted inside the grounds, but no driving around inside was allowed. Every rail to which a dozen or so teams were hitched had a man to look after the horses. He would water and feed them. The feed and nosebags, of course, having been brought from home in the buggy. This attendant collected 25¢ for each team and paid the Fair Association $1 daily for this privilege.

In vogue with society's rich upper crust was the showy pastime of coaching. Occasionally a tally-ho drawn by four horses would wheel through the main gate, its entrance heralded from aloft by bugle. Straight to the Administration Building it would proceed where ten or a dozen ladies and gentlemen, dressed for the part and conscious of privilege, would receive the president's welcome. Very jolly and colorful were these parties of swells, and willing to pose for photographs as crowds gathered to admire in consecutive order the horses, the coach and its occupants.

Early Ferris wheel at the Danbury Fair with the first Administration Building, which burned down in 1897. *Danbury Museum and Historical Society*

Bicycles figured so prominently as transportation during this period that checking facilities were prepared for them too. We are inclined to regard the bicycle as an appurtenance of newsboys and smaller fry, but in the 1890s whole families took to their wheels. So universal was the craze that legislation was invoked to regulate their speed and to preserve the safety of pedestrians. Fines were established for "scorching,"[32] for riding on sidewalks and for coasting down hills.

A fraternity grew up among wheelmen such as later existed among owners of early automobiles. Their conversation was a tangle of chains and sprockets, coaster brakes and tandem rigs.

Now a bicycle is a handy thing for a day's outing, and fashion rode tandem with utility. From Bridgeport and Norwalk, from Stamford and Greenwich, young people and those not so young came singly, in pairs and in parties. Jaunty costumes were noted, among them that of a young woman "wearing a black derby hat, black knickerbockers and riding a diamond frame wheel." Her companion was a youth dapper in "a knickerbocker suit of loud plaid, sporting sea-serpent golf stockings."

I never saw a sea-serpent golf stocking. I hope I never see them, but if they could beat the sport shirts of today that city people deem appropriate for country wear, they must have indeed been astounding.

In 1894, Wheelmens' Day opened the Fair. The Columbia racing team of Hartford attracted much attention for their costumes, light blue tights with the club color in red across the breast. They also "wore the national colors in the form of a sash."[33] Prizes aggregating $1,000 in value attracted some of the fastest wheelers in the country.

The use of a new device by the announcer of these races was noted and commented upon. It was a "megaphone, which renders his words distinctly audible in every part of the grandstand."[34] This was probably a rank exaggeration, but values are comparative.

An announcer broadcasting with a megaphone, 1919.
Frank Baisley

When bicycling was at the summit of its popularity tandem wheels with two or even three seats were fairly common. For purposes of advertising, the Orient company built a giant cycle proclaimed as the Biggest Bike in the World. It was named the Oriten since it had seats for ten riders. "The machine was twenty-three feet eight inches long and its gearing was a cause of wonder to all beholders. The front sprocket and chain were of the size seen on an ordinary tandem; the next wheel back and the chain which ran to the third wheel were somewhat larger and they kept increasing in size until the rear sprocket wheel was over a foot in diameter; the chain which ran on the back

32 This may have referred to slamming on the brakes to leave skid marks behind.

33 *The Danbury News*, October 3, 1894

34 Ibid.

wheel was of great size and strength to stand the strain of propelling the giant bike. Directions for riding came with it. First the wheel was to be mounted by four riders who propelled it until they became proficient; then another man was allowed to mount and the crew was increased one man at a time until it was fully manned."[35]

When it was shown at Danbury in 1898, crowds stood about this remarkable machine all day wondering at its massive tires and heavy gear. By bringing the Biggest Bike to Danbury, the Orient Company made its name familiar to wheelmen and secured valuable newspaper publicity.

For about ten years, the bicycle ranked next to the horse as a grandstand attraction.

Incidentally, two important contributions were made to the speed of the trotting horse by that unlikely contraption scorned by horsemen. At fairs where bicycle races shared honors with harness races, horsemen began to notice the advantages of pneumatic tires. They also noticed that the wheelmen used every known contrivance to break air resistance, which was the principal check on speed. It became apparent to many that the old high-wheeled, iron-rimmed sulky with the driver riding up high where he presented his body to the pressure of the wind, ought to be streamlined, but it took a Yankee, Sterling Elliott by name, to build the first low-wheeled, rubber-tired sulky, which was to cut as much as four seconds off trotting and pacing records. It was first tried at Mystic Park, Boston, in 1900 and within a few weeks came into active and general use.

While the horse and the bicycle were comfortably basking in the warmth of public regard, a new invention was being exhibited at fairs. Perhaps Mazeppa, the educated horse with that gift of prophecy common among literate equines, shuddered for their common future when the first horseless carriage arrived upon the scene.

1899 saw the first use of the automobile as a conveyance for patrons of the Fair.

> *J. B. Cornwall of Bridgeport came to the Danbury Fair today on a Locomobile. He arrived in town at ten o'clock this morning having made the trip from Bridgeport in two and one half hours. He came by the way of Greenfield Hill, which is the longest route between this city and Bridgeport, and quite hilly, in order to test his horseless carriage in hill climbing. The distance by this route is about twenty-nine miles. The power which drives the Locomobile is steam, generated by gasoline. It makes no noise and the only indication that there is any power within the carriage is a little puff of steam which escapes from an exhaust in the rear every few seconds.*[36]

Alas and alack! That little puff of steam!

35 *The Danbury News*, October 6, 1898

36 *The Danbury News*, October 6, 1899

How short would be the time before the self-propelled vehicle was to displace both horse and bicycle, and make even the railroad excursion a thing of the past.

By 1922, the premium list, after a recital of the timetables of the various trains into Danbury, remarked that "the Danbury Fair Grounds is now reached by state roads from all directions, north, south, east and west. Most of its patrons are now coming by auto. Large tracts of land affording additional parking space have been added to the ground, and every effort will be made to cater to the comfort and convenience of the autoist."

As the automobile became more accessible to the average consumer, the infield parking lot was often filled to capacity, 1920s. *Frank Baisley*

6

FAKERS AND THE MORAL TIDE

Those who rented space at fairs for sale of goods or entertainment were known as "fakers" or "pitchmen." They were called fakers because the goods they sold frequently turned out to be either of little value or not as represented after the Fair was over and the merchant had gone, and pitchmen from the expression "to pitch a tent" because a pitch was the name given to the stand of a street merchant or performer.

In spite of the unpleasant implication of their name, fakers might be and often were simply vendors of merchandise, and honest ones at that, like the cider faker and the popcorn faker. On the other hand, the hypnotist and the fortune-teller, the operators of the coochie shows and gambling games were also fakers. They contributed life, color and sin to that important central avenue known as the midway. They might set up an establishment of many square feet under canvas in an area that was rented by the square foot, or the pitch might consist only of a walking permit to sell balloons.

The Midway was the perfect place for concessionaires to sell their wares, and young and old alike were not immune to their charm. *E.J. Doran.*

One midway celebrity of the horse-and-buggy days was a red-bearded Frenchman from Massachusetts whose real name was Boutellier, but who was far more widely known as "the Old Whip Man."

He wore a broad-brimmed hat and a black coat hung smock-like to just above the knees of his short, energetic figure. The two patch pockets of his coat were unusually large and deep, while across his fancy vest was draped a watch chain of heavy gold.

His whips were contained in a sort of knapsack with a carrying strap slung over one shoulder, much as a golf bag is carried. His stock ranged from light carriage whips with lashes of bright colors to the long flexible rods ending in rawhide loops used for driving oxen.

He would select a whip at random and start cracking it expertly, a procedure that never failed to attract bystanders. The small boys within earshot came running at once and formed the nucleus for a fast-growing semicircle of folks who were drawn by curiosity, but who lingered to listen, laugh and become customers.

In the morning, the asking price was $1 per whip, but as the day wore on bonuses were given to stimulate sales, the closing price in late afternoon sometimes falling as low as five for a dollar. Farmers and owners of stables who were acquainted with this sales routine waited from one Fair Week to another to buy their entire supply of whips from this engaging hawker, as year after year the pretensions of his big coat pockets were justified.

Today's fakers differ little from shopkeepers with more permanent quarters. Most of them are glad of a normal profit, and those who might lean toward larceny are kept in line by a Fair management that is known to be intolerant of funny business.

Most are small businessmen who travel from fair to fair in trailers containing neat housekeeping arrangements and a family, including a child or two whose training, likely as not, is guided by the writings of Yale's Dr. Arnold Gesell.[37]

---

37 As director of the Yale Clinic of Child Development, Gesell observed and recorded the changes

It was not always thus.

There is no reasonable doubt that many of these tenants used to be thieves and rascals. One year, the *New York Herald* reported the presence of two rival New York gangs running games of chance on the grounds. All the fakers, even the most respectable, were united in a common aspiration, namely, to make a clean-up Fair Week.

Wherever money was being loosely handled in many small transactions, there used to be pickpockets. Even now, it is a mark of discretion not to carry a wallet in one's back pants' pocket.

These characters did not belong exclusively to the gaslight era, but their numbers are diminishing for it takes time to master the highly-skilled trade of the dipper.

Today, when craftsmanship is being sacrificed to production and the apprentice system has gone by the board, it is simpler to follow the horses or even to get a job with the Department of Internal Revenue.

T. C. Millard, Chief of the Danbury Fair Police force for years, says that most of the experts are known in police circles where their photographs and methods are recorded.

He tells a story of how one busy Saturday, complaints were received in his office of repeated losses of money and jewelry. Mr. Vreeland, a vice president of the Fair, was inclined to discount the theory that a pickpocket was at work, but Mr. Millard instructed his men to keep a sharp look-out. Sure enough, within an hour a well-known operator was brought before him on the lawn of the police barracks. The state police were notified by phone. In the meantime, some of the Fair officials in roguish mood thought it would be good clean fun to play a trick on their friend Mr. Vreeland.

These jokers inquired of the captive whether he thought he could inconspicuously possess himself of Mr. Vreeland's diamond stick-pin. Eager to please as he was, this artiste affirmed that with the help of a newspaper the project might be accomplished.

The newspaper was promptly forthcoming.

Then still in handcuffs, but holding the paper unfolded before him, he walked toward Mr. Vreeland who stood on the nearby plaza talking with some cronies. As if immersed in the news, he stumbled into the vice president and slightly unbalanced him.

With a murmured apology for his carelessness, he then slowly circled back to where stood the fascinated sponsors of his roguery. In one of his palms lay the diamond stick-pin.

Mr. Millard placed this in the protective custody of the lost-and-found

---

in child growth and development from infancy through adolescence, and established the gradients of growth that defined the norms or typical behaviors of children throughout childhood.

department, whence it was redeemed by Mr. Vreeland, who came in quite a lot later to report his loss. At that time, considerable hilarity ensued and the drinks were on Mr. Vreeland.

By 1893, all applicants for space were required to sign the following agreement:

> *In consideration of being allowed to offer for sale our goods at an agreed license or rent at the Danbury Agricultural Fair we hereby promise the Danbury Agricultural Society not to sell or offer for sale any intoxicating liquor or beer, or any article of any description by means of gambling or game of chance, either by offering or placing money or other valuables on our game or goods, and if we fail to keep our agreement, we hereby authorize the said Agricultural Society to cancel our license and to retain any money we paid for said license and forfeit our stand.*

In 1894, the Agricultural Fairs of Connecticut were blamed for an infection of gambling that swept the state.

Lottery tickets were handed out in restaurants entitling the holder of the lucky number to two free meals. So-called "guess" tickets were handed out in theatres and a watch was set going in a jeweler's window. The person who guessed closest to the hour, the minute and the second when it would stop, won the watch. One furniture dealer spent Fair Week sending up balloons to which were attached tickets entitling the winner to various pieces of household goods. A frantic hunt for balloons took place with, as one critic put it, "our leading citizens steeplechasing it across the countryside."

Some newspapers hinted that gambling privileges were privately sold by the management of the Fair itself under cover of the concessionaires' signed promise not to engage in games of chance. That charge was never proved, but the officers do appear to have been guilty of intermittent myopia.

Probably the most famous of ancient flimflams was the notorious "thimblerig," or shell game, that sleight-of-hand swindle in which a small pellet, the size of a pea, and three walnut shells are used—the victim betting as to which shell conceals the object. This fraud often existed in connection with some ostensibly legal operation and was protected by a watchful confederate who kept an eye out for the police. The small folding board, three shells and a pea were easily pocketed when the word "slough" was given. The victim was not so commonly a farmer as might be supposed. More often it was a shrewd-appearing businessman or a well-dressed mechanic who thought he was slick enough to beat a slicker at his own game.

There were cappers for this and other gambles. A "capper" wore quiet clothes and started the come-on by winning first various small amounts, then a substantial sum. It was said in the New York papers that in 1895 native Danbury cappers could be hired at $3 a day to rope in their friends and acquaintances.

That was the year the fakers really ran riot, supplying the gullible investors with every luxury life has to offer, from a cheap alum diamond to a costly and painful experience.

The most profitable pitches were the "envelope" and "picture" games, which were similar. A sum of money, \$20 or \$50, was written on the face of an envelope or on a picture, that amount to be paid to the smart aleck who could pick it out from others mixed with it. It cost 50¢ to play this game at which nobody but the cappers ever won. When a man had lost some money, he was followed by a capper. If he showed signs of going to the police, the capper immediately made overtures to him to return the money.

Another charge leveled at the midway was that such shows as The Nautch Girls, The Streets of Paris and the Hoochie Coochie dancers were indecent. The "muscular" dance was not then generally accepted as it is today by café society, but was labeled "For Men Only" and advertised by its barker as "something which cannot be described." A feature of these performances was the "after show," at which for an additional dime or quarter, "the ultimate will be disclosed." Tall teenagers who felt guilty posing as men were suckers for the after show.

"You goin' in?" one lanky youth would ask another.

"Gosh, yes, I'm goin' to see this thing through if it costs fifteen cents more."

When they emerged from the den of iniquity into the light of day, they had little to say, which led to the suspicion that they had not witnessed

Whether it was in the Streets of Paris, above, or the Streets of Cairo, "For Men Only" dance troupes excited the crowds. *Frank Baisley*

what they expected to see. Those who paid to see five French girls in living pictures saw one draped model in five poorly executed poses.

On one occasion, Isaac W. Ives, one of the Fairs' original promoters and a director of the Agricultural Society, was in high dudgeon over being ordered off the track with his tandem team to make room for Mr. Rundle's horses. Smarting from this snub, he wrote a letter to *The Danbury Dispatch* in which he angrily called public attention to the "demoralizing effects of the most degrading shows of woman's nakedness which the writer saw himself (and for this purpose); and within a few feet of the grandstand four naked women in the most disgraceful attitudes and acts."

In rebuttal of Mr. Ives, an early columnist writing for *The Danbury News* said:

> *Now I didn't hear of anybody who actually saw the things suggested by the steerers but Ike Ives. Ike had better luck than I did, for he owns up that he found a place where they was naked women--and he says he seen 'em his own self. But Ike ain't as young as he was when me an' him was boys together, and his eyesight ain't so good nor so reliable as it was then. Now I've got a son away at boarding school. He was here to the Fair and he went into the same tent where Ike saw the naked women, and he reported that they were giving tablooxes38 there in flesh-colored tights just as they do in any theatre.*

Ike was not alone, however, in his indignation, for another wrathful letter was published signed only by "Indignant Citizen."

> *Is it possible that the miserable orgies enacted last week at the Fairground were chaperoned by the president of our oldest bank, as president; the mayor of our city and a prominent officer in one of our churches, as secretary; and the president of one of our savings banks and elder in the First Congregational Church, as treasurer? Do they expect us to let our wives and daughters go to see their public exhibition of harlots, and our sons to be fleeced by gamblers? Never have the blushes of shame been so thick on the cheeks of honest women!*

The *Republican Standard of Bridgeport* charged that liquor was being sold openly.

> *A few weak attempts have been made to stop it, but they have been only so half-hearted as to result in failure. It was impossible to find a lawyer in Danbury to prosecute cases against the violators, so Attorney E. O. Hull of Chamberlin and Hull of this city, was assigned the duty by the commission.*

There were complaints, too, that betting on the races was being freely transacted at a certain spot under the elms that bordered the track.

---

38 Meant to be "tableaus."

All in all, it looked very much to newspaper readers as if the Danbury Fair was bound for hell in a handbasket.

In the judgment of the Agricultural Society itself, the situation was getting just a bit out of hand and receiving too much unfavorable publicity. A clean-up campaign was begun the next season with the fakers under close scrutiny and supervision.

The first step was to banish even cider and hop-beer from the grounds.

Then the Fair police notified the manager of the "living pictures" that the women, who were already clad in tights, must put on skirts. The manager simultaneously began to pull down his tent and demand a refund.

A young woman trainer of Wallace, the famous lion of the Hagenbeck circus, was discovered by a policeman to be exposing her lower limbs. Hagenbeck promptly cut the act, which, nevertheless, was in his contract and had to be paid for.

A fortune-teller was closed up under an obsolete law so old as to sound like something out of the scriptures, "Thou shalt not suffer a witch to live."

Like a wheel of fortune in a reverse spin, Fair policy shifted from anything goes to nothing goes. It began to look as if the Fair slogan for 1896 was "The Management Strikes Back."

The wave of morality was most unpopular and there went up cries—nay, howls, of "church fair" and "Sunday School picnic." Letters to the newspapers protested the absence of the gay midway of former years, deploring the Fair's narrow-minded conduct.

As a result of all the hullabaloo, a happier balance was established thereafter, and morals were not an issue again until 1899 when, at the instigation of "some reputable citizens," Secretary Thrasher of the Connecticut Law and Order League came to take official cognizance of reported wickedness at the Fair. He proved himself to have been felicitously named.

Bettors were always eager for an unhindered view of the faces, often hanging on the rail for the best line of sight, 1934. *Frank Baisley*

The Law and Order League was an arm of law enforcement holding a charter from the state by legislative action in 1895 before the creation of the state police. Its purpose was "to organize righteousness against organized lawlessness." It was supported by voluntary contributions, acting only when called upon by a body of reputable citizens.

The League's activities were likely to be regarded as an intrusion by members of any local police department. In this connection, Judge Lyman D. Brewster said, "It is only reasonable to expect that home resources be exhausted before outside forces are called in. If the president of the United States should send troops into any state without consulting the governor, there would be a great to-do. A certain amount of jealousy is natural."

On Thursday of Fair Week, Thrasher merely observed, but on Friday he took steps. According to his statement, he "closed fourteen stands where liquor was being sold, and fifteen or more gambling schemes were turned away." In these proceedings, he was rendered no assistance by the Danbury Police, who together with Mayor G. Mortimer Rundle and all other members of the Fair society regarded the clean-up with jaundiced eye.

Instead, to his annoyance, Mr. Thrasher himself was tailed about the grounds by Fair police. Tall, broad-shouldered and blond, he was conspicuous as he sternly stalked past a Fair policeman named Michael Cunningham. Mike, who had stopped to pass the time of day with a friend, remarked to his companion, "There goes Thrasher."

Thrasher, sick and tired of being an object of derision, turned angrily upon Cunningham, who was not in uniform. Cunningham showed his badge, but it was affixed to his suspenders where only a part of it was visible. Thrasher jerked open the cop's vest, the better to see it. Angered by this indignity, Cunningham fended off the secretary by grasping his arm.

At this point, the two men were in a fury, but remembering just in time that they were law-enforcement officers, they repaired to the fairgrounds police station, where Thrasher entered a complaint of assault against Cunningham. Cunningham countered by a similar charge against Thrasher. Chief of Fair Police Bevans decided to put them both under arrest. They were then taken down to the Danbury Police Station and turned over to Capt. Ginty, head of the town police.

Capt. Edward Ginty, a man of action, who had squabbled with one of Thrasher's deputies in the main tent on Thursday, was in no mood to pour oil on troubled waters. He promptly locked up the two of them.

The Rev. Dr. Bowdish of the Methodist church, doubtless feeling some responsibility in the matter, furnished $25 bail for Mr. Thrasher and the secretary was given an opportunity to expound his wrongs from the pulpits of both Methodist and Baptist Churches on Sunday.

On Monday, he called upon Prosecuting Attorney Booth to ask for a warrant for the arrest of Capt. Ginty on the grounds that Ginty had kicked an officer of the league, Otto Burns, after Burns had called Ginty a vile name in the Thursday fracas.

Attorney Booth did not think justice would be served by arresting Capt. Ginty and stated that, "while there might be a technical case against him, the whole affair is so trivial in nature that in the interest of good order and the dignity of the law it is better dropped."

Thrasher announced darkly that he should proceed against Ginty in another way.

This tempest was reported daily by newspapers throughout Connecticut. Charge followed counter-charge. Many letters pro and con were written and printed. Thrasher was compared to Dr. Parkhurst,[39] who was then crusading in New York. The word "Thrasherism" was coined to describe unwelcome censorship and the influence of the League became noticeably less.

Of Capt. Ginty, the *Waterbury Democrat* said:

> *One Capt. Ginty, rough and ready as he is, can do more good for the State of Connecticut than a score of Thrashers. Furthermore, he has the good sense to keep out of the pulpit and attend strictly to the work for which he is trained.*

The consensus was that a meddler had received his just comeuppance.

A postcard depicting the Danbury Fair Police at entrance to the Fair, 1930s. *Frank Baisley*

39 Rev. Dr. Charles Henry Parkhurst (1842-1933), American Presbyterian clergyman famous for his attacks on political corruption and organized vice (*The Reader's Encyclopedia*, 1953 ed., pg 822)

7

WOMAN'S PLACE

No doubt life in Connecticut eighty-five years ago had much to recommend it. The air was cleaner, the streets were quieter and prices were lower, but from a feminine viewpoint, mine anyway, the existing era of the gadget is just dandy.

Those were the good old days when the homes of Danbury were heated by stoves, lighted by kerosene lamps and swept by brooms; when laundry was scrubbed in tubs on washboards and hung out in the sun, or rain, to dry; when cakes were beaten with a mixing spoon, a hundred strokes after the addition of each egg. In stores, there were no ready-to-wear departments, although tailors were plentiful and sewing machines had already been available for twenty years or more. They were in fact manufactured here from 1865 to 1874 by the Bartram and Fanton Sewing Machine Company.

Plumbing was neither complicated nor costly since ownership of an indoor toilet was an isolated exception to the general rule, no very definite connection having been established at that time between the outhouse and typhoid fever. It was 1891 before the first outfall sewer was constructed in Danbury.

Refrigeration was procured by putting a card in the window that read "ICE." In due time, a stalwart with tongs would thump a dripping chunk down into the icebox and track up the kitchen floor.

Although a few militant spinsters had started a campaign for the right to vote for women, their married sisters, for the most part, sought creative expression in their husband's homes, which they busily adorned with crocheted lambrequins[40] and hand-painted china, with ferneries and rubber plants.

Outside the scene of her drudgery, a woman's place was on a pedestal and traces of chivalry were everywhere to be noted.

In the cars of the Danbury Horse Railway, gentlemen bumped each other in their eagerness to offer their seats to any female of any age, a courtesy which ladies accepted as their just due.

Emulating the social customs of New York City, Danbury considered it proper, even smart, for ladies suitably escorted to attend the races. There at the Danbury Pleasure Park, they twirled rosy parasols beneath the summer sun, bowing right and left to their acquaintance. Gentlemen inquired of them, "Is smoking offensive to you, Madam?" and did not pause for an answer before tossing a fresh lit cigar far out across the greensward.

Ornament that she was as she walked aboard on the arm of her natural protector, the versatile creature became a competent housewife as she re-entered her front door, behind which a growing swarm with enormous appetites usually awaited her return.

For her, on October 2, 1888, was published the *Danbury Fair Cook Book*, a clothbound volume of sixty-three pages, "prepared for presentation to the patrons of Danbury Fair by the merchants of Danbury." Sandwiched between pages of "receipts" were advertisements, presumably those of the merchants who contributed to the Cook Book's publication.

Homemakers who depend on cake mixes will be impressed by the lavish use of eggs and lofty disregard of detail in this formula:

GOLDEN CAKE

1 lb. flour, 1 lb. sugar, 3/4 lb. butter, yolks of 14 eggs, one teaspoon of soda, the juice of one lemon.

No directions for mixing and baking!

---

40 Small pieces of decorative draperies hung over top of a door/window or draped from a shelf/mantelpiece.

Suppose the modern housewife does contrive to measure the flour, "a pint's a pound the world around," and that she knows enough to cream the butter and sugar, can this smarty also test the oven temperature by inserting her hand for a second therein? That's how grandma did it. Then, even as the cake comes golden from the oven and cools upon the rack, rendering up its lemony perfume, she's going to have to make an angel cake to use up the fourteen whites. More power to her!

After this recipe is appropriately inserted the full page advertisement of H.M. Robinson, a poultry raiser at 87 Deer Hill Avenue, breeder of the Autocrat Strain of light Brahmas, large-size persistent layers, almost non-sitters. He cites their record:

From 12 Hens in July 200 eggs
From 10 Hens in August 184 eggs
Confined in a small yard and having laid constantly from the previous December.
Chicks at 6 months weigh from 7 to 8 lbs.

Not a bad idea to keep a few persistent layers in the backyard with eggs ranging from 12-15¢ a dozen. But let us return to our cooking.

LEMON PIE - VERY NICE

3 cupfuls pulverized sugar, ¾ cupful of butter, 10 eggs; beat the whites separately; 4 large lemons; grate the outside of the peel, adding the juice, butter and sugar rubbed together; then add the yolks, then the whites of the eggs, then the lemon juice, then the grated lemon peel; then bake with a striped crust.

Sounds delicious, but who has a pie-plate these days to fit this outsize concoction? It must have been baked in a milk pan.

Now for some hearty food.

Conditioned as most of us are by the Old English motif on Christmas cards, why not try roasting a young pig for Sunday dinner.

BAKED OR ROAST PIG

A pig for roasting should be small and fat. Take out the inwards, and cut off the first joint of the feet; boil until tender, then chop them. Prepare a dressing of bread soaked soft, the water squeezed out and the bread mashed fine, season it with salt, pepper and sweet herbs, add a little butter, and fill the pig with the dressing. Rub a little butter on the outside of the pig to prevent its blistering. Bake or roast it from 2 ½ - 3 hours, with a little water in the pan. When cooked, take out a little of the dressing and gravy from the pan, mix it with the chopped inwards and the feet, put in a little butter, pepper and salt, and use this for a sauce on the pig; expose the pig to the open air 2 or 3 minutes before it is put on the table to make it crispy.

Not a word about the big red apple in the pig's mouth, but that ought to be the easiest part of it. The first step is what sticks me.

Next, follows the announcement of a comeback:

BURNED OUT!
But Not Discouraged
The old established house of
Joseph T. Bates & Co.
LUMBER MERCHANTS

Phoenix-like they have "risen from the ashes" of the Great Canal Street Blaze and are now fully prepared to meet every want of the building trade.

Good for Mr. Bates!

Sanctioned by Agricultural society wives, the Danbury Fair became the social event inaugurating the fall season, and preparations were not lightly made. It marked the advent of the new fall dress and suit, of the new hat, gloves and footwear. Wearing of the latter generally proved to be a mistake, hurtful alike to shoes and owner, and *The Danbury News* editorialized against it.

Women all over town painstakingly readied their entries for competition in the department in which they excelled. Bread, pies and cakes had to be fresh-baked, preserves and pickles brought up-cellar, and fancy work wrapped in tissue paper.

In 1887, a dozen classes of premiums were dedicated to the domestic arts.

Class 12 was for bread, dairy products, preserves and pickles, but hard and soft soap and beeswax were also included here. Class 13 was for cakes and pies "made by any person not in the baking business." There were ten classes in the Ladies Industrial division for display of items of home manufacture. Most of the rugs and quilts were of fine workmanship; some indeed have survived as heirlooms that may be as proudly shown today. But the pillow shams and fire screens, the cotton and worsted crocheted tidies, the embroidered illuminated mottoes, the hair work, the beadwork, the shell work and the waxwork have passed into limbo along with the thread and needle case, the comb case, the slipper case, the glove case, the watch case, the hair receiver and the whole amazing array of things to put things in.

Mrs. Martilla Jenkins has a friendship quilt that she says was originally exhibited by her grandmother in 1888, adding that its fascination was such for her that, as a little girl, she was always willing to make the bed that it adorned.

Each block was composed of many tiny pieces carefully cut and fitted about a central square on which is embroidered the name of its maker, the whole constituting a perfect fusion of sentiment with utility.

Imagine having forty friends willing to contribute to such an undertaking. I wish I had such a memento. I can picture myself lying late of a morning in tender recall of the embroidered past, mulling over the careers of the worthy women who were once my dearest friends.

"That sloppy-looking square was pieced by Isobel Foster, she that was Isobel Macomber. I'm surprised she did it at all, not the fancy work type. She was sent home from three boarding schools and one college for climbing out of dormitory windows, but she's made a perfectly elegant wife and mother—just what she had in mind all the time, I suppose. And that pink and blue square with the neat stitches, Daisy Crooker did that. She was a lady if there ever was one, and what did it get her? Four lumpy, loud-mouthed offspring that take after their father. Florence Wilson worked a bluebird over there under her name. Florence was slim and blond and wanted to be an actress, but she thickened up and the last I know she was president of a Woman's Club in Wellesley."

Too late now, I suppose, but what a keepsake it would be. I'd even cherish an autograph album.

Beneath all this cookery and patchwork were subversive influences and a new look.

The period was being enlivened by two innovations attributable to the bicycle: the bloomer and the ankle-length skirt. Women had made the discovery that a skirt that swept

Under the Big Top
Top: Gourds displayed in a variety of shapes and sizes, 1920s. *Frank Baisley*
Bottom: Concessions, demonstrations and so much more filled the cavernous space 1976. *Danbury Museum and Historical Society*

the floor was ill-adapted to anything but the ballroom. An abbreviated skirt, known as the "rainsy-daisy," was designed and first worn in New York, whence its pattern spread to the suburbs, the provinces and the hinterlands.

The new freedom was shocking to many who felt that exposure of an ankle subjected the male animal to cruel and unusual temptation and so furnished an occasion of sin. Many males themselves, strangely enough, protested vigorously against the new skirt as immodest, immoral and altogether disgraceful.

I surmise that the men, unconsciously perhaps, disliked the new styles for the insurgency that they symbolized. Amelia Jenks Bloomer, an ardent suffragette,[41] was supposed to have advocated the costume that bore her name, although I did hear Martha Deane say one morning that it was introduced by the Bloomers, an English vaudeville troop.

A writer for *The Danbury News*, although impressed by the new costume, was inclined toward tolerance. He remarked the appearance of a young lady wearing a "tailormade bloomer suit which fitted snugly." At the top, I presume he meant.

"She was so pretty and shapely that there were people on the grandstand when she passed along who found out afterward that they had neglected to score the trot."

Savant, however, writing in *The Daily Dispatch* rebuked his fellow citizens thus:

> *We go to the Fair and see women straddling bicycles and we commend their skill. Some cities advocate the bloomer costume for bike riders, and right here in our own city we permit the abbreviated skirt and endorse it.*

Wives, sisters and daughters, smiling a little at all the to-do, continued nonetheless along their primrose path, cutting up for patterns the very newspapers containing that windy blather of protest.

When it came right down to a choice between a pedestal and a bicycle, all able-bodied femininity took to the bicycle.

Alas, in a few more years there will be left no eyewitnesses to that period known as "Victorian."

But, alas again, while we are at it, let us get practical and mourn for our own generation. In a few more years, there may be no one left to recall the age known as "Atomic."

41 A woman who seeks the right to vote through an organized protest.

8

WAR TIME

Pearl Harbor was bombed on December 7, 1941, and we were at war. Some fairs ran on a reduced schedule for the duration, but the state of Connecticut was one big defense plant. Due to shortages of transportation and manpower for unessential activities, the Danbury Fair remained closed for four years: 1942, 1943, 1944 and 1945.

While the fortunes of the Fair were in this state of suspended animation, the tinkers, the tailors and the candlestick makers in our vicinity were turning out bullets and bombsights. A whole citizenry, united in the defense effort, was rolling along in high gear.

John Leahy was in overdrive.

Anybody who tried to run a fuel oil business during World War II will understand that the problems of buying and distributing the product would have stumped Paul Bunyan. We were lucky to have a water terminal in Norwalk where we could receive in large storage tanks whatever heating oil was allotted us by the Fuel Administration. Tank cars were taken over by the Army and Navy in such numbers that for two whole winters we received most of our range oil or kerosene in 55-gallon steel barrels loaded in boxcars. The labor and expense of handling and unloading the contents of those heavy drums into tank trucks and thence into storage tanks made kerosene a troublesome commodity, but people had to have oil and there was no other way to get it.

During this period, oil dealers who were active and bitter competitors in normal times held daily phone conversations and swapped products, both of which neighborly customs have since fallen into disuse.

At our White Street office, John supervised our principal business of selling heating oils and gasoline, as well as a younger but thriving business in cooking gas and a third business, which, starting as a small machine shop, had grown in war time like Jack's beanstalk until it had succeeded in pushing the trucks out of their garages and across the street, where temporary shelter was prepared.

Rationing, with its books and coupons, was a complication that led to numberless time-consuming interviews and very nearly doubled the clerical and delivery work of marketing oil.

The Jewel Gas, or propane, business was only about five years old at the time. It was my personal pride and diversion. I had helped John organize it, did the clerical work and took care of the floor sales. I must say I knew my pipe fittings.

Once when I was showing Fred Kling, who managed the operation of this department, an order for some such items before I mailed it, he approved my efforts with these words, "It's O.K. Send it right along. You're half a plumber."

While not actually glowing from the warmth of his praise, I did, at the time, complacently interpret the remark as a compliment. I'm pretty sure it was.

A number of the most competent men both in our office and on the trucks were called into service. Many of those left were draft-deferred only by reason of their employment in the machine shop on defense projects, so that the oil business suffered most.

When Norbert Mitchell, our New Milford manager, left for basic training in October 1943 there was nobody in our skeleton organization to take his place.

"Who," pondered John aloud, "can be spared? It will have to be someone willing to drive back and forth, who can get along with the drivers and dispatch the trucks. He will have to check out the gas station and the degree day[42] deliveries. There's isn't anybody left who answers that description."

"It ought not to be hard to learn the degree day system," I commented by way of encouragement.

---

42 A heating oil industry calculation that is used to predict when a customer will need more heating oil based on their average consumption and high and low outdoor temperatures.

"You don't think so?" he replied. "Say, I never thought of it, but you're the one. We can divide up your work here. You're flexible, you know the work and you could do it well. Why, what are we waiting for? You can start going over the accounts right now."

Since I am the kind with whom flattery not only gets you somewhere, but with whom it works like magic, I somewhat reluctantly became "half an oil man" by promotion from Jewel Gas to the position of Branch Manager. At least I had a title for three years, a dubious honor that I soon should have been glad to relinquish along with the fourteen-mile drive each way.

Of a winter morning when the sun rises at seven and shines upon a frosty wonderland of ice-coated twigs and branches and the glassy highway beckons, I am willing to take my chances at 20 mph along with the army of the employed.

At night, I am less resolute.

That first winter, it seemed to get dark and to start sleeting with painful punctuality at about four o'clock in the afternoon. At six, with a salt contraption fastened to my windshield-wiper, I would back out of the warm garage with all the zest of a cat about to plunge into the Housatonic. As the Ford tacked capriciously back and forth across Route 7, the slapping rhythm of its tire chains echoed a familiar metrical form to which the words, "It was the schooner Hesperus that sailed the wintry sea,"[43] grimly fitted themselves. Clasping the wheel with rigid fingers, I sympathized deeply with that unfortunate skipper.

In addition to the abnormal amount of energy involved in keeping our customers warm and within their rations, John also spent time and effort in the shop where war orders were being filled.

My free-wheeling husband is a machinist by trade, having served an apprenticeship while in his teens in Turner's Machine Shop. At the age of twenty, he felt he was wasting his youth and that it was high time to get started on a career. Accordingly, he rented some space in the unused corner of another shop, had some stationery printed, and obtained an order for piston rings from the old Locomobile Company in nearby Bridgeport. Thus, practically overnight, he became an industrialist as the newspapers call everyone in Connecticut who runs a little shop of his own.

By the time oil started to be popular for home heating, he had saved enough to buy a truck. However, there was widespread opinion, circulated no doubt by the coal dealers, that the country's available oil supply would soon be used up and that it was best for consumers to keep their coal grates ready to slip back into their furnaces in a year or two. This instance of cultural lag was on par with the

43 From "The Wreck of the Hesperus" by Henry Wadsworth Longfellow.

putting of whip sockets[44] on early automobiles. But to be on the safe side, John held on to the machine shop and continued to fill orders for existing customers without bothering much to secure new accounts. As time passed, he concentrated increasingly on centerless grinding.

"Centerless grinding" is a finishing operation on round surfaces that must be accurately sized within very close tolerances. A great deal of it is done in this state where so many planes and so much small ammunition is manufactured.

When the war came, orders poured in and our small shop overflowed into two long garages, which were converted by the installation of windows, toilets and fluorescent lights into a sizable little defense plant, containing a battery of forty machines for precision grinding.

We ground tumblers for locks, bolts and studs for airplanes, .30 and .50 caliber bullet cores, parts for surgical instruments, clinical thermometers and the glass pistons for hypodermic syringes.

These varied undertakings would keep an average mortal fairly well occupied, but John is just naturally self-winding. The more he does, the more he becomes interested in doing. I used to protest vaguely that he shouldn't work so hard until I realized that work in the sense of toil is something he rarely undertakes. He enjoys concentrating on a new project the way some people enjoy solving a double-crostic.[45] Perhaps it is relaxing in the same way—who am I to understand the functioning of such eccentrics? I certainly shouldn't want a husband to be unemployed.

John came home from his office on a spring evening during these trying times, sat down in the kitchen and began untying his shoes. He insists on keeping his slippers in the kitchen upon the pretext of efficiency, although how the saving is accomplished has never been clear to me, since after pulling them on, he straightway proceeds upstairs to the bathroom.

"Remind me," he called from the landing, "there's something I want to talk over with you later. Nothing that will affect you, so don't get worried."

John knows that I dread more than anything to be joggled from a good comfortable rut into which I have recently settled. My peace is always being disturbed by some new scheme of his, but not as a rule at home or after dinner, since at six o'clock he sheds the cares and vexations of the business day, just as he parks his shoes beside the washing machine. He remembers where they are, of course, and can put his hand on them if he needs to, but they are temporarily laid aside.

The fireside talks we have about business are not even discussions, but have their value I suspect as a settling-out process by which the trivial sinks to the

44 A socket into which the butt end of a whip is inserted when the whip is not in use.

45 A multi-part word puzzle. The first step is to guess words from provided definitions to enter into columns of numbered dashes. Then you copy those letters into the corresponding numbered squares of a diagram to reveal a famous quotation and author.

bottom, leaving the more important substance clarified and ready to be dealt with the next day. I don't have to say much, but I do have to listen with interest. It is best not to knit or sew at such times.

As usual, he remembered without being reminded.

"Funny thing happened today. I bought a share of Danbury Fair stock."

"You did? Why?" I wondered, politely putting down my knitting.

"Remember Mrs. Prince? The one on North Street? She came in this morning to pay her oil bill. We got talking and she wanted to know if I would buy her share of Fair stock. She's had the certificate in the house since her father died and she said she'd rather have the money than the stock."

"How did you know what it's worth?"

"I didn't, but I called up Mr. Rundle. He said that if the Fair were operating it would be worth $75 or $100. The par value is $25. For the last few years, all the stockholders have been getting are free passes."

"Did he advise you to buy it?"

"No, but he didn't advise me not to. He said there's a big mortgage on the fairgrounds because of the fire, but he thinks they'll get going again when the war is over. I think they will."

"I shouldn't mind having a few shares of Fair stock," he mused. "When I was a boy, the Fair was the biggest thing in town."

"Once a year," I put in.

"It helped support a good many families," he countered. "Mrs. Prince showed me lots of old pictures of the exhibits and grandstand shows. I remember most of the acts. She had a picture of a balloon ascension that I well remember."

I settled back.

"A balloon ascension in those days was a ticklish performance. I'm not talking about captive balloons now. The kind I mean was free. The balloon itself was just a big round bag of silk or cloth, made nonporous and with an open end."

"What was in the bag to make it rise?"

"Hot air," John replied.

"Just hot air? How clever of them to find a use for it! But how did they collect it and get it in the bag?"

He grinned.

"I can tell you how it was done. Shall I?"

I was all attention.

"First, they had to find a suitable spot for the take-off with no trees in the way. Then, a shallow depression was formed by scooping out some earth where the fire was to be. Here stove wood was piled. From this point, a small trench was dug half a dozen feet long and the diameter of a stovepipe. The purpose of the stovepipe was to direct a current of air pumped by a small hand bellows into the firepit from beneath. Do you know what a bellows is?"

"Of course I do. It's one of those air squirters they used to have in blacksmith shops to start up the fire on the forge."

"Right! Well, after kerosene was poured over the wood and the fire started, this current of air caused the heat to rise rapidly toward the mouth of the balloon, which rested on a metal framework previously erected over the fire. At first, the limp sections of balloon fabric had to be held up and away from the flames. There were always plenty of volunteers for this work. All the boys wanted to help."

"You seem to know all about it, yourself," I observed.

It's hard to imagine today, but the earliest balloon ascensions were done without the benefit of a basket for the rider. *Frank Baisley*

"The warm air rose fast and caused the fabric to belly upward. There were four ropes attached to the top of the bag. When the balloon became buoyant, it had to be held down by main strength until the signal came to let go. The volunteers then competed for the job of holding the ropes."

"Did the man stand up in his car and wave his handkerchief?" I wanted to know, recalling pictures I had seen labeled "Early Balloon Ascensions."

"Car? There wasn't any car. The cords that hung from the lower edge of the bag were joined and hooked to an assembly of parachute package above a trapeze seat or bar. The daredevil just sat on this bar. Then he gave the signal to release the four ropes. All the boys had to let go at once."

"How did he steer it or get it back down?"

"It was no dirigible. He went whichever way the wind was blowing and when he wanted to come down, he pulled on a rope. This rope worked a round knife to cut a cord and release the balloon. The parachute was then supposed to open and, if it did, he had no trouble coming down!"

"Sounds risky to me."

"There's a better word for it than that. One man was killed when his parachute failed to open, and a girl jumper who couldn't cut herself loose landed in the trees on Town Hill right at the foot of Prospect Street when the balloon finally collapsed."

"Was she hurt? Did you see it happen?"

"That was before my time, but Jarvis' father saw it. He said her trapeze caught on some fire alarm wires, and she hung there until some of the neighbors brought a ladder long enough to reach her. They found that three strands of the rope had been cut, but either she had lacked the strength to cut through it or the knife was dull. She was able to give reporters an interview that evening."

"Lucky girl! What did people get paid for taking chances like that?"

"A hundred dollars, maybe, but there were other rewards. The jumpers were heroes. They received top-billing and were followed around all day by every kid in town."

"And I suppose they were posthumously awarded a blue ribbon for foolhardiness," I scoffed. "Did you ever talk with one of these daredevils? What was he like?"

"Sure I did," John proudly asserted. "Same type of fellow that drives the cars in thrill shows—I was one of the guys that held the ropes."

"So that is why you want a share of Fair stock?"

"Maybe it is, in a way," admitted John.

"The whole thing looks run down to me," I stated, feeling that the time had come to express an opinion. "Those low buildings along Backus Avenue need painting in the worst way."

"They need more than that. They are sway-backed and their roofs are full of holes. Same with the poultry building and the horse barns, but they could all be made serviceable. It would take money, but the Fair could earn enough to pay for it, a little each year. What the Fair needs is a manager, someone who is interested in planning for the future. Mr. Rundle is getting old now, and this war can't last forever."

"What is the amount of the mortgage?"

"Mr. Rundle didn't say, but I think I'll go downtown and have a talk with him someday soon."

I knew then that Mr. G. Mortimer Rundle, president of the Fair, was going to be surprised at the interest taken by his newest stockholder.

9

RUN AROUND

The next forenoon, with the newly acquired certificate burning a hole in his pocket, John headed for Mr. Rundle's office.

Mr. Rundle, at the age of 88, had been president of the Fair Association for over twenty-five years. Sole successor to his father, Samuel's, interests in hat manufacturing and real estate, G. Mortimer had contrived additional careers for himself in politics and banking. If not universally beloved, as who is in the course of such activities, he had many loyal followers and was a figure of distinction in Danbury as survivor and representative of a bygone era. He projected the dignity of his presence from a height of little over five feet, and his keen brown eyes bespoke untarnished intelligence. Intimates and detractors referred to him as "Morty" or "old Morty Rundle," depending upon viewpoint, but mostly behind his back, which in no way diminished his prestige. His age was in itself impressive.

In his high-posted Victorian home on Farview Avenue, his way of life was that of two generations before. Besides his housekeeper, he employed a trained nurse to supervise his diet and general good health. A gardener-chauffeur worked on the premises and drove him downtown daily, sometimes to attend directors' meetings of the various banks and boards with which he was connected, but regularly to the office of the Danbury Agricultural Society where, Fair or no Fair, he kept regular hours.

To this niche over a Main Street store John came to have his little talk. Advised by a sign on the door, he knocked and entered a dim, narrow room containing an old-fashioned roll-top desk with chairs to match, a tall safe ornamented with gilt scrolls, and several shelves of ledgers. He had not realized until that moment that these quarters adjoined the real estate office of George Nevius, the Fair's secretary and Mr. Rundle's old and closest crony. The connecting door was open. Both officials were in the president's office and seemed slightly startled at the invasion.

"They were up to something," John told me later. "They scooped some stuff into a drawer in a hurry—could have been a checker board or a second mortgage."

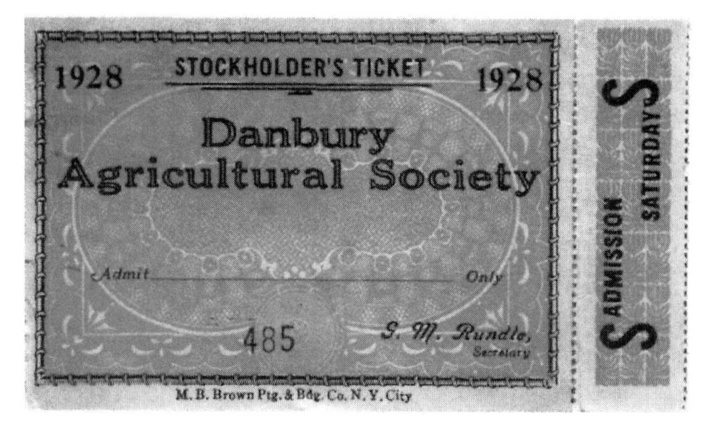

Hands having been shaken and the weather touched upon, John proffered his stock certificate for transfer.

"I paid $100 for this," he remarked. "Was that somewhere near the right price?"

"Well, now." Mr. Nevius answered. "It's a little hard to say what the right price is. There hasn't been much trading in Fair stock lately. I have recorded some transfers, made the last one when Jim Northrup died two years ago, but nobody has sold any stock for all of eight years. What was it those two shares of Barrett's sold for in 1934?"

He looked at the president.

"Seventy-five, and '34 was a good year," was the prompt reply. Mr. Rundle turned to John.

"A hundred was enough. What did you want with it?"

John just told the truth.

"Why, I always liked the Fair," he said.

"Any new stockholder who takes an interest is welcome, although there have been few times in the history of the Fair when its position has been more critical."

John smiled.

"I expected that," he said. "It always happens whenever I join a Chowder and

Marching Society. The candidate admires an organization and is eager to associate himself with such a fine upstanding group. As soon as he becomes a member, the first thing he hears is what tough shape the club is in and how all the members, instead of relaxing, must put their shoulders to the wheel and push it uphill. The Fair has been a success for over seventy years. How can it be on its last legs just as I buy a share of stock?"

"That," Mr. Rundle replied with a smile, "is the difference between the outside and the inside."

The president was very frank about financial matters and seemed to take pleasure in explaining them. He understood the workings of every department for each had in great measure been subject to his supervision. As his conversation skipped nimbly from horse racing to pumpkins and from machinery to needle-work, his eyes spoke his affection for the annual event in which his father had been the leading spirit.

The morning must have been enjoyable for John seemed more interested than ever in the Fair, as he described the interview to me at dinner.

G. Mortimer Rundle.
Danbury Museum and Historical Society

"I stayed until the chauffeur came and for half an hour afterward, while he waited for Mr. Rundle, who had started to tell me about the original stock issue. I never knew how many shares there were. He owns a hundred of them, which is the only large block of stock."

Between soup and salad, as I heard the rest of the story, a suspicion came over me that we were buying a lot of unnecessary bother for $100.

"If it hadn't been for the fire in 1941, Mr. Rundle said the Fair could have weathered the war without any trouble. After the other fires, there had been insurance plus a bank loan, which could be repaid in a few years from income. The overhead, without any repairs or painting, runs high in the four figures, which makes more borrowing necessary, but, to a fair without definite prospect of opening, the bank will no longer be justified in making further loans. If the war should stop before the clock runs down, the Fair could again gradually pay off its indebtedness. The fire was what made the difference."

Both John and I remember that disaster. It had come about as the result of an attempt to supplement the Fair's income from rentals.

In the fall of 1940, the management voted to lease certain space for opening an ice skating arena. The space consisted of the level circular court formed inside the main ring of exhibition booths. This is the area that is covered during Fair Week by an immense canvas umbrella known as the Big Top. In winter, this site was ideal for a rink.

The enclosure was accordingly flooded and skates were purchased by the lessee for rental. Local boys and girls patronized the arena enthusiastically until a cold winter afternoon in January 1941, when that old red devil, Fire, slithered into this Eden of sound administration for a fourth time.

The fourth fire at the Fair demolished the Big Top and Administration Building, 1941.
Danbury Museum and Historical Society

It broke out in the corner of one of the exhibition booths where a small heating stove had been set up so that the patrons could take off their shoes and don their skates in comfort. A defective flue was said to have been the cause of the blaze, which twinkled along the circumference of the arena and rapidly crackled into the tinder of the adjoining main building.

The Fire department made the two miles from the Ives Street firehouse as fast as they could, but even before they reached the scene, fire was breaking through the roofs and firemen had all they could do to prevent the spread of the flames to the poultry building and adjacent stands.

John recalls driving past the fairgrounds at about 1:30 p.m. on the way to Norwalk. Seeing smoke rising from the direction of the Big Top, he supposed the skaters had made a bonfire. When he returned at 5 p.m., the fire was still burning, but there was no doubt about what had happened.

We have a picture taken the morning after. Five steel tent poles of the Big Top stand starkly down the center of the freshly frozen arena ice, encircling which lies a ring of black ruin. The site of the administration offices is symbolically marked by the remains of three empty safes.

Again, there was insurance and once again the Fair turned to the Danbury National Bank for a loan to repair the destruction of its most valuable property. All that summer, carpenters sawed and hammered away to erect its central structures for the third time.

The new main building with its white colonial facade was ideally planned for the purpose it was to serve. Like its predecessors, it was two stories high and was bisected by a tall archway that formed the entrance to the Big Top.

On the ground floor to the left were the offices of the president and secretary, complete with bars and modern plumbing. Here too was a first aid room and additional toilet facilities. Upstairs, was a private dining room with serving pantry.

On the right side, the vice president's suite was similarly equipped and contained a cashier's office, like a bank teller's cage in which to transact business with concessionaires.

At the top of the stairs in this wing was a small room for private negotiation complete with another most intriguing bar built in and concealed behind the woodwork.

Adjoining was the treasurer's office, which did not include a bar of any kind or even a toilet—a plan which I always thought showed good judgment up to a certain point.

All these rooms were walled with knotty pine and furnished with modern desks, tables and office equipment at a total cost of about $35,000. Who was to know that the Fair that opened in 1941 for business as usual would catch the last glimpse of profits for four long years? As things turned out, that period followed when, principally because of restrictions on the use of gasoline, the new offices and exhibition booths stood empty and forlorn while taxes and interest mounted as roofs cried out for repairs.

Impressed by Mr. Rundle's wisdom and encouraged by his friendliness, John formed the habit of dropping into the Fair office. Mr. Nevius, the secretary, was often present and John enjoyed the flow of reminiscence and shrewd observation that marked the conversation of the old friends. The glories and disasters of bygone days were fresher in their recollections than the present-day crisis.

How these Yankees from way back came to trust the Irish upstart who listened and commented so agreeably, I'm not sure. The three were far from kindred spirits, but they did have one common interest. It was the perpetuation of the Danbury Fair.

This community of purpose was no doubt the basis for Mr. Rundle's sale to John later that year of fifty shares of Fair stock, for he certainly was in no need of money then or later. He must have felt the human need of seeing his efforts sympathetically seconded, of enlisting the drive and devotion of a younger man on behalf of the languishing institution that he held dear.

He made no mistake, for John was by then a zealous partisan. He began advertising in the local paper to buy Fair stock and we were busy for many weeks with letters and phone calls from owners of single shares. One by one these were purchased, easily at first from owners who had half-forgotten the yellowed, crumbling certificate, good now only for admission to a Fair that might never again open its gates.

Then came tougher customers.

Some answered the advertisement and, upon finding out what price was offered, declared that $100 was "nowhere near enough money." The mere fact that somebody wanted to buy it gave the stock glamour in the estimation of holders who had never before suspected it of having a market value.

John dealt with these individuals courteously, explaining the situation as it was, stating why he believed the price of $100 to be fair and that he could pay no more. Those who really intended to sell anyway sold. Several of the ones who decided to wait for a better offer, sold a little later when they finally realized that the boom in fair futures was a one-man project.

The most exasperating were the ancestor worshippers, subdivided on the basis of sincerity into sentimentalists and others.

The sentimentalists were good folks who felt a pang at parting with a souvenir of happy childhood hours or with the possession of a departed parent. They were honestly torn between sense and sensibility. That they were toying with the idea of selling was always an indication that they might succumb.

The others really knew what they were about when they carried on for hours about the hallowed past. When the roll is called up yonder it will be a matter of record, I trust, that in 1943 several proper Danburians were guilty of hocus-pocus in an attempt to increase the selling price of their Fair stock.

One prim lady kept John dangling the greater part of the winter and enjoyed more of his society than I did by the simple expedient of asking him to come over and explain a little more about the difficulties of the Fair. Since she was a two-share woman, he always hastened to her side.

After a long cozy chat over cookies and blackberry wine, she invariably concluded, "You're just as nice as you can be, Mr. Leahy, to go to all this trouble to explain things to me and, if I ever decide to sell my stock, I'll certainly let you know. I can't make up my mind to sell right now. Father set such store by the Fair. I don't believe he would want me to."

"How long has your father been gone?"

"It will be thirty years next May."

"He might—if he knew the shape it's in now," John would intimate a little brusquely, reaching for his hat.

Home he would come agreeably bemused, but a little crestfallen. In a week or two, the phone would ring at dinnertime and he would again announce hopefully, "Mrs. Lotsatalk wants me to stop in at her house this evening."

Irked at last by this repeated performance, I retorted snappishly, "For heaven's sake, offer her more money. Isn't your time worth anything?"

He looked a little hurt.

"You think that's what she's up to?"

"Do you think she goes on this way for the pleasure of your company? It's a lot of bother making blackberry wine."

"She's a pleasant lady," John mused, "I never thought of it that way."

"Well, the least you can do is try it and see."

He brought home the certificates that night. Much against his principles, he had paid $125 for each of her two shares.

I chortled in an unladylike way.

The pleasant lady continues, however, to be one of John's staunchest admirers, still calling him up occasionally for advice. The last time she suspected her oil man of double-dealing, she changed to Leahy's Oil Company, a step which constituted some form of reparation, I suppose.

Usually with ancestor worshippers, it was best to outline facts and figures in one friendly lesson and then to relax for a time. Out of some enchanted mailbag, months later, there was likely to come a letter enclosing the familiar rectangle bearing in old-fashioned type the imprint of "The Danbury Farmers and Manufacturers Society."

Let it not be said, although it probably has been, that John was frittering away his time on an amusement enterprise while the country was at war. The time he could spare for picking up stock was strictly limited by the demands of the working day and the fact that he has never been able to break himself of the habit of sleeping soundly for six hours every night.

What he did have in bountiful supply were dogged perseverance and a sense of humor. These enabled him, working in odd moments, evenings and Sundays, to acquire one-by-one a couple hundred shares, many of which were purchased by mail from heirs who had moved away from Connecticut. We had to hunt up addresses and write scores of letters, often enclosing with the offer a check for $100. Sometimes, the stock would be forthcoming by return post. Other times, the check would be sent back without comment. Many of letters were returned stamped "Moved - No Forwarding Address." Our mail was full of surprises.

"When I get through with this," I proclaimed one day, "I am going to apply for a job with the Bureau of Missing Persons. They'll be glad to have me."

"When you get through with this," John quickly rejoined, "I'll get you a job as treasurer of the Fair. That way, you won't have to leave home."

"You'd better think that over," I warned him. My incompetence with little numbers has been demonstrated time and again. It has been noted and commented upon by many of my co-workers and franker acquaintances.

"I've *been* thinking it over. You can't add, of course, but nobody ever got a cent away from you."

"Of all things," I quavered, my voice rising emotionally.

"Don't worry about it—yet," he said. "We'll get you an adding machine."

It came to John one day as he was filing away a newly acquired certificate and counting his hoard that he must be close to owning half the stock, a state of affairs he had never dreamed possible at the outset. One share more than half would give him a controlling interest and place the management in his hands.

This result was worth a special effort. He determined to ask Mr. Rundle to sell his remaining fifty shares.

"What do you think the chances are?" he asked as he started off on this dubious mission.

"It's about like asking him to part with his mustache, but go right ahead. I don't believe he's the sentimental type."

With these few words of encouragement ringing in his ears, John set out for an evening at the home of the president. From this interview, he returned intact and reported that Mr. Rundle did not appear much upset by his request. To be sure, he didn't give in immediately, but after a few months of familiarity with the idea, he developed a liking for it.

In my opinion, he was neither overwhelmed by John's energy nor worn down by his persistence. In the light with which time illumines the motives of human behavior, I think now that what Mr. Rundle wanted was the simple survival of the Fair to which so great a part of his life had been given. Since he had no son, he finally selected John as most likely to succeed in its management. In the end, he sold his second block of stock as if sentiment had never existed, stipulating only that the large oil painting of "Quartermaster" be returned to his home, since it was his personal property.

We were the sentimental ones as we regretfully took down and wrapped the portrait of the mettlesome black stallion.

Nothing looked right in its place until 1948, the year of the Fair's Diamond Jubilee.

Then *True* magazine reproduced a watercolor by Warren Baumgartner entitled "Oxen Pull at the Danbury Fair" as a cover for its September issue.

A pair of twin Holstein oxen owned by Winthrop Smith were portrayed pulling a stoneboat against a background of blue sky, rolling meadows, a Ferris wheel and a typically spruce-looking crowd of onlookers. Above and over all, Old Glory rides the autumn breezes.

On Governor's Day, *True* generously made presentation of the original to John so that now Quartermaster's spot in the president's office is occupied by another painting that captures the feeling and flavor of the Fair.

It will be a reminder and a token to us in our time as the older picture was to Mr. Rundle in his.

Mr. Rundle retained the presidency during his lifetime, and in that capacity happily attended every Fair including the one of October 1950, which shortly preceded his death in the same month.

On left, artist Warren Baumgartner presents John W. Leahy with the original painting "Oxen Pull at the Danbury Fair," cover of *True* magazine. Connecticut Governor James C Shannon on right.

10

INVENTORY

At a stockholders meeting in 1943, John was elected general manager and at last he had a free hand to spend all his time and as much money as he could scrape together on the Fair. To be sure, he owned little more than half the stock, but the joys and headaches of administration were his alone.

"Beware your heart's desire," I misquoted at him, "for that you will surely get."

He understood what the author and I had in mind, but was far from being disconcerted by our cynicism.

"When the war is over," he maintained, "we'll have the best and the biggest Fair those grounds have ever seen."

"If we should live that long," I said to myself.

"Get a big pad and a pencil," he instructed me, "and let's get the nuts out of the shells."

Here was the situation as we found and recorded it.

The fairgrounds extended over an area of 154 acres, 28 of which were in swampland through which the outlet of Lake Kenosia flowed in a narrow channel. There was from this source a plentiful water supply for general purposes. Five dug wells furnished drinking water.

The inventory of buildings in good condition included the new combination Administration Building and exhibition area, a concrete and steel grandstand, two large concrete cattle barns, a spacious machinery hall and a two and a half-story farmhouse where the superintendent, Henry Johnson, lived. This house and the big barn that went with it had been kept in repair because they were in regular use. Machinery Hall, rented for two years to the Bassick Company for defense work, actually had been earning $187 a month.

"Can you think of anything else that doesn't need money spent on it?" I asked. "This isn't a very long list."

"Put down location," said John.

"Location?" I echoed.

"Yes, nobody traveling north or south on Route 7 can fail to notice the grounds. As you look down from the highway, the big sign on the horse barns, 'The Great Danbury Fair,' hits you right in the eye. The sight and recollection of it have more advertising value than a whole row of billboards."

"Sounds reasonable, what else?"

"Well," he concluded, "there's good will. The Fair has been running for over seventy years and is known from the Atlantic Ocean to the Hudson River and beyond. You can't laugh that off. It has a name and standing as a well-conducted fair. While there's no way of measuring the cash value of good will, it is worth more than the grandstand. I know that."

The other side of the picture was not so cheery.

Fixed expenses came first.

"Leave spaces after each item, we'll fill in the figures later," John directed. "Put down taxes, electricity, telephone, interest on mortgage, insurance—"

"Not so fast," I protested, scribbling furiously.

"Insurance," he repeated, "fire and liability, Main Street office rent, superintendent's salary, truck and car expenses. Got those? All right. Necessary repairs come next."

We explored the condition of the older structures and found it strictly unsound. The horse barns needed new roofs. The great rambling poultry house needed roof, sills and other extensive repairs. The oxen, swine, sheep and goat barns needed what amounted to complete restoration. Sills, framework, siding and roofs just hung together more from old established custom than for any other reason and were ready to crumple beneath the weight of a good orthodox New England snowfall. They all needed paint.

"Now add roads."

"What's the matter with the roads?"

"There aren't any. The midways are thick with dust in dry weather and soupy with mud when it rains. The grounds should have some decent roads. And while you're at it, put down trees."

"Some of the old elms look as if they're dying."

"Some are, but others are being choked by poison ivy. All the trees need trimming and spraying right away, but many of them can be saved. We must plant more as soon as we can."

"All this," I sighed, "is going to cost a lot of money."

"What a head for business you have." John exclaimed, laughing and one-handedly exploring my skull for signs of overall enlargement. Then, kissing me soundly as if to repair past omissions, he added, "But that's not what I like best about you."

"No, I know better than that. It's my flexibility," I retorted.

Early in 1944, the Connecticut Aberdeen Angus Association made application to rent space for its first annual show and sale. This activity was one for which the premises were well-adapted and appropriate. Besides, every little bit of income helped. For the event, their committee chose Machinery Hall, largest of the exhibition buildings and recently vacated by the Bassick Company.

Beginning at 10 a.m. on Monday, May 22, seven bulls and thirty-nine females of this handsome, hornless breed were presented and judged by W. Alan McGregor, president of the National Association. They were real Scottish aristocracy from the famous bloodlines of the Blackbirds and Blackcaps, the Ericas, the Eisas, the Barbaras and the Prides. When groomed for the show ring, these low, black, blocky, well-proportioned beef cattle looked exactly as if they were upholstered in gleaming black plush. They had won more grand championships at the Chicago International Stock Show than all other breeds combined.

In the afternoon, the sale was conducted by Roy G. Johnston of Bolton, Missouri. To us New Englanders, unaccustomed to the chant and technique of the Midwestern auctioneer, it was a revelation of skill and showmanship. All the cattle were sold at prices that seemed at the time fantastically high and when, at the climax of one of his professional chants, Mr. Johnston obtained the price of $800 for Bandocap's Lulu, owned by the University of Connecticut, the whole stand burst simultaneously into applause.

The annual sale was held at the fairgrounds for four years until breeders in the eastern part of the state petitioned for its removal to a spot more centrally located. John and I fondly remember it as the first bit of real fairgrounds activity in which we had a part. It was also a brief, but happy, experiment in the rental of facilities.

The prime duty of a manager is to find out where the money is coming from. What the Fair needed was a transfusion of dollars, those red corpuscles in the lifeblood of any healthy business enterprise. John applied himself to finding a donor, but money is hard to borrow in wartime for a peacetime project. The Danbury National Bank held a first mortgage of $27,500, which was teetering on the verge of foreclosure. Additional funds were urgently needed to pay overdue taxes, cover outstanding bills and defray running expenses.

To keep the Fair solvent, John therefore decided to put $25,000 of his own money into the corporation, this loan to be secured by a second mortgage.

The way that money evaporated was amazing. It took care of the most pressing obligations, but was not enough to get more than a start on the work of saving the buildings from further deterioration.

By the summer 1944, although the end of the war was not exactly in sight, it was plainly indicated after V-E Day, and since he was anxious to get repairs well under way during warm weather, John made a further investment in the Fair of about $18,000.

Every weekday morning that summer, he took his lunch in a paper bag and set out for the fairgrounds in his veteran Ford coupe with a toolbox over the rear bumper. By dint of classified advertising and the local grapevine, he flushed from cover a collection of carpenters, painters, roofers, road builders and fence makers, many of whom had retired from the practice of their crafts five to ten years earlier and who constituted, temperately speaking, an altogether remarkable crew.

Among these conscripts was a dearth of muscle power and no great unity of method or purpose, but a tractor with a hoist was purchased to provide the former and, before long, an *entente cordiale* having sprung up from nowhere, the carpenters began to hammer, the painters began to paint, and the roofers began to roof with a vengeance. Things got done that season.

John was in his glory working outdoors in the hot sunshine, frequently declaiming that he never felt better.

This was the year he began to get acquainted with the regular population of the fairgrounds and found communion with Nature most rewarding.

He had almost always been able to tell an English sparrow from a robin and a bluebird from a crow, but there his ornithological perception came to a limping halt. He now began to notice with astonishment and interest the variety of birds that nested confidently in the trees, fields and vacant buildings.

These idle acres formed a natural sanctuary where goldfinches and other warblers sang from fences and red-winged blackbirds swung on the tall grasses of the marsh. In the early summer, whole tribes of young swifts and barn-swallows lined up on the telephone wires where flying lessons were endlessly conducted. There would be glimpses of an indigo bunting[46] on a mullein[47] stalk or the rarer flash of a purple finch. Sometimes, a young mourning dove, tame as any pigeon, would walk across the lawn. Behind the farmhouse a white lilac bush was reserved each year by a pair of orioles. On warm evenings, nighthawks swooped low overhead in pursuit of moths.

Families of killdeer abounded, laying their eggs most imprudently all through the hayfields. When John found a nest with eggs, he would drive a tall stake to warn off the mowing machine. Every day, he used to pass a nest on the very edge of the roadway without causing the slightest concern to the mother bird who quietly sat out her time until her young ones were hatched and running about in the funny teeter-totter manner of sandpipers.

With the approach of fall, two great blue herons came and stood about on one leg in the shallow swamp water, peering down their long bills for fish. This was also the time of the annual grackle invasion. They always arrived in force, whirring, wheeling and chattering in hoarse voices, then coming by unanimous consent to rest briefly in a clump of tall elms. The branches of the old trees buckled beneath the weight of their thousands like giant blackberry bushes overloaded with iridescent fruit.

There were foxes and squirrels, possums and woodchucks, snakes and turtles, and countless families of brown bunnies with white tails. Not all of these made daily appearances, but the ones that did delighted John hugely. He had never been a farm boy and so discovered the world of small wild animals rather late in life. They charmed him. He gave names to the regulars: Freckles was a red fox. Mixture was a red and gray fox. Big Joe was a black snake whose life was repeatedly saved by John's intervention. That summer, while I was in Maine visiting my mother, he wrote me about these creatures in his telegraphic style.

> *Yesterday morning while I was hurrying around north end of Big Top, Mixture and I met head on. We were both so surprised we just stopped and stared at each other. Then Mixture trotted leisurely off toward the swamp. Today I saw Freckles walking up center of road from airport gate. First I thought a dog was coming along. Freckles saw me on the porch in front of the office, but he kept coming. Cantered past me straight into his den under the Big Top. Seemed very friendly.*
>
> *Found a fat lazy black snake 4½ feet long, lives down near the pump house. I visit him every few days where he lies out full-length sunning himself, and warn him not to go out in the field and get cut up in the mowing machine, for he's a beautiful specimen. I call him Big Joe.*

---

46 A small seed-eating bird in the family *Cardinalidae*.

47 An herb plant of the figwort family with woolly leaves and tall spikes of yellow flowers.

And at other times:

> *The playful children of Mr. and Mrs. Woodchuck are fun to see every time I drive along Church Row. A family lives under nearly every stand. The young ones scamper for their holes when they see me coming, but peek out and smile as I drive by. They are getting so used to me now that sometimes they don't bother to hide.*
>
> *There's a possum around the grounds. Works mostly nights. He makes me mad when he digs little holes in the grass out in front of the grandstand. Looking for grubs, I suppose. His wife and family live under the lunch counter near Machinery Hall. The family is shy, but Pa and I are pretty well acquainted by now. Could almost have picked him up by the tail a few nights ago. Didn't quite dare and he didn't quite trust me. I told him he is welcome to dig the grass anywhere but in front of the grandstand.*

Ever since that summer, we have been steady patrons of the Bronx Zoo. When we go there in the spring and fall, we take along our small relatives, Susan and Jackie, as a cover-up. But on a recent trip to the Tampa Fair, we took time out to explore the Fairmount Park Zoo in Philadelphia, the National Zoological Park in Washington, and viewed en route at least three other smaller collections of animals without the benefit of Jackie and Susan, although I must admit I felt guilty. We always visit the winter quarters of the circus in Sarasota, we peer through the port holes at Marineland, and we explore snake farms, ostrich farms and bird sanctuaries. Florida must be glad to see us coming.

11

WAYS AND MEANS

By late September, both John and the fairgrounds were improved in appearance. Stronger and leaner, John had turned healthily brown, and most of the buildings, likewise reinforced, had changed from bilious saffron to glistening white.

While there were more fundamental and extensive improvements, nothing so altered the aspect of the whole property as did painting. Spotted about the meadows, their green roofs hardly distinguishable amid the general verdancy, the offices, halls, barns and the sheds looked as if some super-housewife had just given them a thorough scrubbing and spread them out on the grass to dry.

There were a few exceptions.

On Church Row, the restaurants of the various Protestant denominations, each owned by its respective congregation, stood amicably side-by-side. Over the years, people had become accustomed to patronizing the eating place of their religious preference, but even hard-shelled Baptists, who had a stand of their own, could be found on stools along the counter of the Swedish Church, whose specialty was oysters. The Church of the Disciples produced chicken pies notable for large chunks of white meat in rich gravy. The Methodists, Universalists and Episcopalians had each become specialists in the concoction of some delicacy, their masterpieces ranging from clam chowder to roast beef and from turkey sandwiches to baked Virginia ham. With little or no advertising, the booths always did a rushing business, as there arose upon the clear October air what was certainly the true odor of sanctity.

Church Row, 1930s. The multi-storied building at the top of the photo is the original farmhouse, which was the caretaker's residence and later the summer home of the Leahys. *Frank Baisley*

Tastes in outside decoration, however, had differed here even more variously than creeds.

Since it occupied a prominent and strategic location starting from the north end of the grandstand and running opposite the Big Top, John wanted this avenue to look like the rest of the fairgrounds. In his haste to get on with his work, he thought it best to establish conformity among the sects by first anointing all the stands liberally with white paint, and later getting permission to do it from the numerous church committees, who would have to consult at length and vote to accept the change along with the donation of labor and paint.

He acted accordingly, but when he finally did get around to approach the various Men's Leagues and Clubs, the Ladies' Aid Societies, Guilds and Associations, he found to his dismay that although each was friendly and appreciative of the painting, not one of the organizations felt it could offer any hope of reopening its eating stand.

"Many of the older members who used to cook and serve the food are gone," one chairman explained, "and there doesn't seem to be the enthusiasm or the time to spare among the younger people. They are willing to contribute money or do some baking for an occasional church supper, but nobody wants to assume the responsibility of management, and not many are free to work regular shifts at the counter for a week."

Even the Council of Catholic Women at the other end of the grounds no longer felt up to the sustained effort of operating their large restaurant, but preferred the smaller income that could be realized from its rental.

"Times have changed all right," philosophized John to whom these decisions were a disappointment. "Nobody wants to work anymore."

"There are a few left," I pointedly insinuated, regarding his mud-caked shoes and trouser cuffs, "Don't give up."

But John took seriously the defection of Church Row and perhaps began to wonder for the first time whether the public taste for country fairs had remained constant in this changing world. We knew the church stands would be missed.

Perhaps it was just as well that colder weather put a stop to these delightfully constructive endeavors as our bank balance was down again to a low two-figures.

At a stockholders meeting on January 6, 1945, I became the duly elected treasurer, thus achieving a new high in flexibility, but since John's office is competently staffed with bookkeepers, it was my fixed intention to regard this position as honorary.

At a later meeting, however, one of the five stockholders who put in an appearance arose and, eyeing me severely, requested, "Will the treasurer read the report of the Fair's financial condition?"

"Who? Me?" I felt like saying and looked indignantly at John as the author of my embarrassment. I could just as readily have obliged with the Lithuanian national anthem, but John was equal to the occasion.

"We have not prepared a formal report," he answered for me, "but I can tell you the way matters stand." He did it in such a friendly and competent way I felt quite proud of him. He is no bookkeeper either.

That night he came home with a folder full of papers with little numbers all over them and some correspondence, which I found to contain the most amazing jargon. One was the copy of a letter that the bank had written in reply to an inquiry about the Fair's credit:

> *The subject has been a depositor for many years, balances averaging medium in three to medium low in four figures. They are indebted to us at the present time for low in five figures. No accommodation has recently been requested, and we should have no hesitance in recommending them to you for the amount mentioned in your letter, ($19.50).*

"Couldn't they have just said, "O.K. for $19.50?" I wondered.

The funny part about the figures was that I really knew quite a bit about them already in a general way. For accuracy, which I have since found needful in certain instances, I have come to depend upon the adding machine, which with no brain at all is my superior at arithmetic.

For the benefit of those to whom bookkeeping is not abhorrent, I am including a statement of expenses made, not by me, for the previous month of November.

FIXED EXPENSES - NO REPAIRS OR UPKEEP

			PER MONTH
Interest on First Mortgage Nov. 1, 1944			
$27,500		$1,100.00 yr.	$91.67
Interest on Second Mortgage			
$25,000		$1,250.00 yr.	$104.17
Interest on Note			
$18,257		$912.85 yr.	$76.07
Interest on Additional Loan Through Feb. 15, 1945			
Current Pay Roll	$891.93		
Purchases	$1,926.41		
Interest at 5% on	$2,818.34 =	$140.92 or	$11.75
Town Taxes (Actual)		$2,916.68 yr.	$243.06
Fire Insurance (Actual)		$3,032.88 yr.	$252.74
Liability Insurance on Property		$250.00 yr.	$20.83
Main St. Office Rent		$25.00	
Office Telephone		$6.70	
Office Lights (Estimated)		$2.00	
Superintendent's Salary		$28.85 week	$115.40
" House Light (Estimated)		$7.00	
" Phone		$3.25	
Truck & Car Expenses, License, Repairs etc. (Est.)			$15.00
Gasoline, Tires, Oil (Estimated)		$200.00 yr.	$16.67
			$991.31
State Income Tax (1944) Actual		$163.26 yr.	$13.61
Federal Excise Tax		$13.25 yr.	$1.10
FIXED EXPENSES ABOUT $250.00 a WEEK OR PER MONTH			$1,006.02

It looks simple and easy enough to understand, but a lot of little numbers had to be written down and added up to arrive at the conclusion that it cost $12,000 a year for that sleeping beauty of a fairground just to lie still, not even batting an eyelash.

The most pressing expense was the obligation to the Danbury National Bank. In the good old days, if the interest on a mortgage was paid when due, nobody had to worry about making payments on the principal. In fact, a loan was considered a good investment if interest at 4% was regularly forthcoming, but some bureaucrat dreamed up the system known as "amortization" as a by-product of the Federal Housing Authority—a term signifying that a portion of the original loan itself must be paid back yearly. This procedure frequently involves borrowing more money from another source, which also becomes subject to amortization. True, it is a nuisance, but the practice is supposed to have a strengthening effect upon the financial and economic fabric of the country and, if so, I am in favor of it.

The letter that follows came as no surprise.

DANBURY NATIONAL BANK
Oldest Bank in Western Connecticut
Danbury, Connecticut

November 13, 1944

Mr. John W. Leahy, General Manager
Danbury Farmers & Manufacturers Co.
Danbury, Connecticut

Dear Mr. Leahy,

This is to notify you that this bank holds a $27,500.00 first mortgage on the property of Danbury Farmers and Manufacturers Company and on November 1, 1944, there should have been a $2,500.00 amortization payment as well as the quarterly interest payment of $250.00 on January 31, 1945, for which the regular notice has previously been mailed to you. As yet, these payments have not been received and we are therefore obliged to inform you that your mortgage is now in default. Naturally, such a status cannot be permitted to remain for any length of time. Will you please therefore see to it that payment is made forthwith.

Yours very truly,
Charles Jost,
President

"'Forthwith,' it says," John muttered.

"That means 'without delay,' 'directly,'" I interpreted.

"Why don't they say how much delay they'll stand for?" he quibbled.

There was just about enough cash on hand together with the proceeds from the sale of some useless farm machinery and equipment to stave off the oldest bank in western Connecticut for another three months. Mr. Jost was most sympathetic and cooperative, but banks just do not choose to go into the fairgrounds business. They have troubles of their own.

Late in February 1945, there was a heavy wet snowfall. John was always uneasy about the weight of snow on the broad, gently sloping roofs of the larger buildings and had it shoveled off as soon as possible, but this particular storm began at about noon and lasted all night. In the morning, Henry Johnson called up to say that the roof of the poultry building had collapsed.

This was a double crusher. John dashed to the scene to find the situation as represented, only more so. The roof and wall timbers were tangled together on the snowy ground like a pile of jackstraws.[48] Of the structure, which had been 60 feet wide by 150 feet long, about one third remained standing.

Carpenters were hastily rounded up to save what was left by bracing and closing it in. The work of restoring it would have to wait until spring and at a conservative estimate would cost $8,000 or $9,000.

The loss was aggravated by the fact that during the past summer this large exhibition hall had been remodeled from sills to ridgepole. The roof had been re-tarred, the siding inside and out had been painted, new window glass had been set, and the benches and wire cages put back complete, ready for the poultry show to open.

With our bank balance now standing in the low two-figures, $15.34, to be precise, John turned his attention to the problem of retrenchment.

For a long time, the fairgrounds had been worked as a farm by the superintendent, Henry Johnson, who lived with his family in the roomy dwelling at the north end, the same house occupied by Nolan at the time of the first big fire. He kept some cows and chickens and planted a garden for his own use, receiving a salary for his services in maintaining the trotting track, cutting the hay, and overseeing the premises. The Fair Association owned three horses, which did the grading, plowing, mowing and hauling.

Normally, the bulk of the hay crop was consumed by the exhibition cattle during Fair Week, but there was a great quantity of loose hay on hand in all the mows that had accumulated over three seasons.

John had watched Henry the previous spring as he hauled loads of manure away from the long pile behind his barn and spread it over the fields.

"What's the idea behind this manure spreading?" he would demand of me every time his nostrils were offended on a rare day in April.

---

48 A game in which a set of straws or thin strips (called "jackstraws") is let fall in a heap with each player in turn trying to remove one at a time without disturbing the rest

John knew the answer to that question very well. All he meant by asking it was that he would like to review the pros and cons of the economics of farming on the fairgrounds. I am supposed to be an authority on farming since one of my grandfathers had a farm in Maine where I spent all my summers up to the advanced age of thirteen.

Obligingly I rehearsed the classic cycle as set forth by my Grandfather Cole.

"You keep cows to make manure to raise hay to feed the cows."

"I feel foolish doing that," said John. "Let's find out what it would cost to hire the hay cut and baled. It would be easier to store and handle, easier to sell if we had any left over. Right now there isn't anywhere to put another year's crop."

A study of the little numbers representing the cost of running the farm showed that it would be more economical to sell the livestock and discontinue farming, at least for a while.

Sorrowfully, John broke the news to Henry, who promptly got a better job on an estate in Redding and moved his family straight to it. It didn't matter much to him.

That left nobody to work on the half-mile horse track, which mattered a lot to a few horsemen who still used the track for training purposes.

For many years, it was the established practice for trainers of trotting horses to move in bag and baggage to the fairgrounds of their choice and make use of the facilities. The horse barns and half-mile track at Danbury had always been made available to horsemen for training purposes long before and after the trotting season. Some horses were quartered here throughout the year.

One of the many horse barns at the Danbury Fair, with the dirt trotting track on the right.
Danbury Museum and Historical Society

There was a nominal monthly charge of $3 a stall for stable rent, but aside from being hard to collect, the revenue never came close to paying a caretaker and maintaining a track, which required the constant attention of two men with hoes and rakes to keep the weeds out and the surface smooth.

This service, however, was supposed to foster good will among the owners who benefited thereby, and to secure a good field for the trotting races.

It sometimes worked the other way for there were always prima donnas among the trainers who would up and leave on some slight provocation. When they weren't feuding with each other, they were apt to gang up on the management.

What with the cost of maintenance, the replacement of small tools that broke and disappeared as if by magic, and the need for constant supervision, keeping the fairgrounds open all year for this purpose was far from profitable, but it had continued at Danbury as it did at other fairgrounds even after World War II put horse racing in the category of unessential activities.

John was hesitant to break with a custom by which the Fair had become the traditional sponsor of trotting, but when considerations of expense were added to the risk of fire occasioned by the year-round occupancy of the barns, he decided to close the fairgrounds entirely for the duration.

When he broke the news of this decision to the few remaining horsemen, they agreed albeit somewhat reluctantly to co-operate.

There was one die-hard, however, by the name of O'Hara.

When John told this trainer, who still enjoyed squatter's privileges here with his trailer and two horses, that the Fair could no longer afford the luxury of keeping up the track, he flew into a rage.

John patiently explained the difficulties the Fair was having and why the grounds were being closed.

"Now," he concluded, "you can see what it costs to provide facilities for a few horses to train here. Until the Fair is operating again, there is no way for us to make up the loss."

"We pay you stable rent," growled O'Hara.

"Some of you pay $3 a month. We are still trying to collect hundreds of dollars from a lot that didn't pay."

"You Irish son of a bachelor," shouted O'Hara. "I never liked you anyway!"

"No need to get mad," soothed John. "Just think things over. What does the Fair get out of it?"

O'Hara pondered for half a minute, then came up with the correct answer.

"Why, you get all the manure," he vouchsafed.

"I understand that," said John. "It's not enough."

12

DOODLEBUGS AND WATERBUGS

It is the seldom-realized dream of every fair association to derive some income from its property during the idle fifty weeks of the year. Upkeep will devour surplus far sooner than most people suppose and a big plant like the fairgrounds needs all the income it can produce.

In May 1940, the directors of the Fair had voted to lease the grandstand and office facilities, parking area and space inside the trotting track to Stuart McLean, then vice president of the Agricultural Society, for the purpose of conducting midget motor car races. Although this was actually an intra-family arrangement, it was not lightly entered upon.

By the terms of the lease, Mr. McLean undertook to build a one-fifth mile macadam[49] track inside the half-mile clay horse track complete with proper fencing

49 A road surface made of compressed layers of small broken stones, especially one that is bound together with tar or asphalt.

and lighting. He would also pay for all costs of upkeep including the expenses of water, lights and toilets. He committed to carry liability and compensation insurance and assume responsibility for any damage to the property, including "any broken window glass."

The rental was moderate, averaging little more than $100 a race the first year, after which it was to rise to $150 "if the lessor decides to permit races during the calendar years 1941 and 1942."

The track was built according to prescription and the venture prospered for two summers. Races were held, weather permitting, once weekly from May through October with twelve to sixteen meets a season. Mr. McLean must have replaced any broken glass, for all went merry as a marriage bell until Pearl Harbor halted spectator sports.

Midget auto racing in its early days was a strictly amateur sport with an interesting origin.

In the period of rapid development of the automobile that followed World War I workers were irresistibly attracted to the centers of its manufacture. Schools for mechanics blossomed in every populated area and a worship of speed set the feet of young men itching to step on the accelerator of a racing car.

There were two serious obstacles to the ambitions of these lads. Generally speaking, a racer was too expensive for private ownership. Automotive engineers for large companies, as well as commercial suppliers of gasoline and motor oil, have long used the Indianapolis Speedway as a laboratory, but for every car so sponsored there were scores of qualified drivers. Suitable tracks were likewise scarce because of their cost.

Not to be denied, however, youthful devotees of internal combustion from Los Angeles to Kansas City to Long Island began working almost simultaneously in shops and garages to build small-scale copies of the big racing cars. They worked nights and Sundays, separately and together, every man of them his own engineer and mechanic, until homemade midget cars were racing against each other on ball fields, horse tracks or wherever there was a patch of level ground.

Somewhere I have read that what amazed the Europeans most about the Americans in World War II was that every American soldier could drive a car, and almost every one could repair it when it broke down. This example of the "American Way" is more impressive to me than the billboard representation of a plump sedan filled with a prosperous looking family obviously stuffed with Sunday dinner.

Variously powered by outboard, motorcycle or Ford motors, the midgets increased in numbers and public favor on tracks of various sizes until by a sort of unanimous agreement the hard-surfaced one-fifth mile oval became standard.

In 1931, the Miller Automobile Company of California began to manufacture a factory-built midget engineered by the foremost designer of racing cars, Gerald Offenhauser. The price of an "Offy" was high, $18,000 or thereabout, but these new cars were so powerful that they were able to pay for themselves by their higher winnings of prize money.

When it became evident that the Offys were returning a profit, money for their purchase was more readily forthcoming and many well-known drivers formed partnerships with owners on a share-the-earnings basis.

Up until the time the Offenhauser made its appearance, chances had been fairly equal among the contestants. Indeed midget auto racing had qualified as a true sport by providing contests involving individual skill and physical prowess. However, a good driver in an Offy was hard to beat except by another good driver similarly mounted.

Later this inequality was to prove disruptive, but in 1940 when the midget races first started in Danbury, there were not enough of the newcomers to excite anything but admiration and envy.

By that year, the sport had increased in popularity to the extent that several recognized weekly papers catered to midget fans in an idiom peculiarly their own.

The little racers themselves were affectionately called "doodlebugs," their followers were "bugs," and the winning driver of the yearly nationwide popularity poll was named "King Doodlebug." The track was "the Oval," a driver was a "lead-foot" or "speed-merchant," an accident might be a "spill," a "spin," a "flip," a "crash," a "tangle" or a "pile-up." Hitting the fence was "rapping the lumber." A rainy night was a "wash-out," and a decrepit old car was a "junker."[50] The operator of a speedway was never referred to as the owner or a manager, but always as the "promoter," a designation which to John's ear has such a fly-by-night connotation that he always rears away from its application to himself.

After the Japanese surrender on Sunday, September 2, 1945, the government issued an edict to lift gasoline rationing. John, under the guidance of Mr. McLean and with the assistance of Mr. Jarvis, decided

Stuart McLean.
Danbury Museum and Historical Society

50 It's interesting to note that many of these terms are still in common use today in the automotive world.

to try a few Saturday nights of racing before cold weather.

It took them about three days to make their arrangements, but it had taken me only one to head for Maine in the Ford.

I arrived there on Tuesday and was considerably surprised to receive a telephone call from John on Wednesday evening.

"You'll have to start back in the morning," he instructed me. "We're opening the midget races Saturday night."

"Why, I just got here," I objected. "I'm not unpacked yet."

"Fine, that's great," exclaimed John. "Glad I called tonight. Saved you all that bother."

"What on earth," I protested, "am I supposed to do at midget races? I never saw one!"

Some nuance of irritability in my tone grated on my husband's sensitive ear.

"Listen," he advised me. "I hate to do this to you, but you're the treasurer of the Fair. Catch on? If you want to keep the job, you'll have to look after the money. Just come home. You'll find out what to do when you get here."

The crackle in this pronouncement betokened great excitement in Danbury and brought me up short. I hastily recanted.

"Sorry! I didn't understand. Hold the job open. I'll see you tomorrow night."

Back in Danbury, my first duty Friday morning was to exchange a Danbury Fair check for two very heavy bags of change. This was for our six ticket sellers, each of whose metal boxes contained $300 in silver and small bills along with

The midget auto races always kept the crowds on the edge of their seats, 1947.

serial-numbered roll tickets in two colors.

The admission price was $1.20, 20¢ of which would go to our silent partner, the Department of Internal Revenue. There had been no time to have reserved seat tickets printed, but there was a half-price child's ticket. Free parking and clean toilets were included.

I studied the check-out sheets and found them easy to understand. Mr. Jarvis, who had always assisted Mr. McLean in this routine, would be on call to help me. Nothing to it, I decided.

On Saturday morning, September 8, the many-colored midgets began to arrive in the pits for speed trials. They were towed on specially built trailers and were accompanied by their racing teams. A team consists of a driver, a mechanic and two pushers, since racers are geared too high to start on their own power.

The purpose of the speed trials, which are held on each track at the start of a racing season, is to determine the relative capacities of the cars, thus enabling the handicapper to enter each in its own class.

The handicapper is the master of the midgets. His word is law in all disputes between the drivers themselves and between drivers and management. He is neither a member of the racing club nor a representative of the promoter, but rather a steward in whose hands rests responsibility for fair play and upon whose calm judgment and firm tact many ticklish decisions depend.

Our handicapper was "Cappy" Lane assisted by "Lamee" Crovat. Both were officials at Madison Square Garden where they had learned quite a bit about both the presentation of sports events and the handling of the hot-headed young participants therein.

Before 6 p.m. that opening night, customers were turning into the grounds and by 7 p.m. one parking lot was filled.

A racing card is ordinarily composed of eight events. First come three races called "heats," ten laps each in which the three top drivers qualify for the first and second fifteen-lap semifinals that follow. After the semifinals, there is a class B race for those cars that fail to qualify in the first three heats, and after that a Consolation Race for the hard-luck boys and the pilots of junkers. The two top drivers in the "B" and the "Consy" are also eligible for the feature race of twenty-five laps. This main event comes at the end of the card with top money going to its winner. The size of the "field," or number of contestants, may vary the above procedure somewhat, but twelve or fourteen cars in the "main" make an exciting race.

To a racing fan, it is almost as important to see and hear the warm-ups of the various entries as to be present at the races. The voice of each motor, unrestrained by a muffler, has a rhythm all its own that is recognizable on nights when the wind is right over two miles away on Main Street, reverberating against the giant sounding board formed by the grandstand.

There the trained ear of a listening fan will catch the cadence and he will proclaim, "That's Tony warming up. Come on, Joe. Let's go!"

From 7 to 8:30 p.m., a steady procession of automobiles entered and came to rest on the green meadows of the fairgrounds. Many cars, conforming to the tastes of their young owners, were accessory-laden and tastefully embellished with raccoon tails, as were the motorcycles that spluttered to their resting place alongside the pony barn. But there were also newer, more conservative cars, containing older yet-still-enthusiastic occupants.

The crowd was in holiday mood. The war was over, its trials were done, and folks had money in their pockets, which at last could purchase entertainment and the gasoline to reach it.

We watched their arrival from the upstairs windows of the treasurer's office. Racing with whoops to place first at the ticket windows were groups of teen-agers. There were foresighted middle-aged people carrying cushions and blankets, and there were lightly-clad young couples hand-in-hand, happily oblivious to weather and temperature. A boy in a wheelchair was pushed along by his buddies and a pretty girl on crutches was making fast time between two escorts.

The three ticket booths to the north, center and south of the grandstand were each manned by two sellers, all exceptionally qualified to give fast service. Two were cashiers in business offices and four were tellers from the local banks. Lines moved briskly at their windows until 6,700 spectators had found seats on the grandstand and bleachers.

Even in the warm-ups, the roar of the midgets is exciting and provocative. Like a call to arms, it quickens the pulses and stirs the emotions of susceptible hearers, among whom I suddenly discovered I belonged. That sound, combined with intermittent rounds of applause from the stands, brought me to the mistaken conclusion in no time that my services would not be required in the office until intermission.

Above: The Mighty Midgets on a third-mile track with the starters looking on from their stand, mid-1950s.
Opposite: Midgets race into the first turn of the fifth-mile track, mid-1950s.

Hastily closing the safe and twirling its combination, I scurried across the plaza to the grandstand.

The voice of Nat Kleinfield was coming over the loud speakers as he introduced the drivers.

Nat is a top-flight announcer. During five seasons of racing in Danbury, he saved many a situation by his fast-thinking and fluency.

When the races were late starting because of a wet track, he amused the waiting throng with colorful, human interest stories while automobiles drove round and round on the oval to dry it off. When there was a pile-up and horrified spectators started to rush out on the track, he talked them out of this dangerous venture and back to the safety of their seats. When lightning struck a pole and followed the wires into the grandstand where smoke and the odor of burning insulation were diffused, he quelled the early stages of a panic with his ready explanation.

In the rapid narration of a race, his voice thrillingly communicated his own sincerity and a depth of feeling strong enough to kindle in many a first-nighter the bright intense spirit of the racing fan. The second race was getting underway as I slipped into an usher's seat near the top of the stairs.

Batteries of lights from the two towers and the roof of the grandstand were focused on the white track and the diagonally red-striped white fences that enclosed it. On the bright green grass of the infield waited two wrecking cars and a shiny ambulance. Wide-eyed, I watched the midgets as they lined up before the starter, Doug Clark, waiting for the signal. There were Schindler, Ritter and Wofsy, Delzio, Tatro, Fennelli, Duffy and Lindberg. I can see them all now, as I later came to know them.

Wide-eyed I soon ceased to be, after Johnny Ritter spun off the track in the second lap, and I stopped peeking between my fingers as Len Wofsy's motor failed and the cars of oncoming drivers missed him by a hair's breadth. Schindler's win in the #4 Caruso Offy was only a roaring in my ears as the head usher tapped me on the shoulder the next moment and motioned me to follow him down the stairs.

"The sellers want more change at the south booth, Mrs. Leahy. They sent a cop up to bring it back, but he couldn't find you. Your office is open, though," he remarked, as if in passing. Bill was always the gentleman.

I made haste to my post of duty. Sure enough, Captain Downs was on the steps.

"They're still coming in at the south end, Mrs. Leahy. The boys want two hundred ones and five rolls of halves."

Breathlessly, I fumbled at the combination of the huge black safe. It was new to me and I kept turning the dial past the stop. The critical surveillance of the officer added to my confusion as I tussled ineffectually with the uncooperative beast.

"Women!" I interpreted the officer's inner thoughts. "Women haven't got the sense God gave geese! What good is a woman on a job like this? Doesn't know enough to stay in the office. Goes off and leaves the doors unlocked. Now she can't open the safe."

For the fourth time, I set the combination carefully. It hadn't worked before, but I was recovering from my semi-swoon. In desperation I braced myself and gave the handle a steady pull. The door opened with a big sucking sound like a cow pulling her foot out of the mud.

"Just fits tight, I guess," volunteered the captain, who suddenly appeared more friendly. He took the change and was gone.

"Mrs. Leahy," came a voice from the lower office, "there are three gentlemen down here who want to see you about refunds."

Downstairs the "gentlemen" quickly closed in on me.

"We want our money back," one of them demanded roughly.

"What's the matter?" I inquired.

"We come to see Nazaruk race and he ain't here. You advertised him and he ain't here. You can't get away with that. We're gonna get our money back."

I had overheard some conversation relating to Mike.

"He is still in the Army," I replied, "at some camp on City Island. He races nights when he can get time off from duty, but tonight he called up to say he couldn't make it. Don't you want to see the races?"

"Naw, we come to see Nazaruk and he ain't here. We want our money back."

This struck me as a bit peculiar.

"May I see your ticket-stubs?" I requested. They had them.

"And your return checks," I added.

"What return checks?"

"The ones you received at the pass-out gate when you left the grandstand."

"Nobody give us no return checks. Why ain't our stubs enough? They show we bought our tickets. You can't toss us around. You give us our money back."

I temporized.

"I have no instructions about making refunds. This is our opening night. If you will wait a few minutes until he comes over between races, Mr. Jarvis will take care of you. He is the Assistant Manager."

"Here he comes now," I was relieved to add as the screen door opened.

"Mr. Jarvis, these gentlemen want refunds, but do not have their return checks."

Mr. Jarvis was courtesy itself, but he made short work of my problem.

"Yes, gentlemen," he beamed at them. "Now by which gate did you leave the grandstand?"

"The north end."

"Just step over there with me," he said. "I think we can straighten this out."

Very few people leave the stand between the first and second heats and the pass-out man was sure he had given checks to all of them. Furthermore, he remembered these applicants for refunds as the first ones to have made their exit at the end of the first race. They had been given return checks.

"Now, boys," Mr. Jarvis told them, "if you want to use your return checks to see the races, get them up right now. If not I'll have the police escort you off the grounds."

They looked at each other and decided to produce the necessary pasteboards. Unabashed, they handed them to the gate-man.

"Jeez, but you're tough here!" one of these individuals loudly remarked to Mr. Jarvis. The boys had been planning to get their money back on the stubs and then to use the return checks for reentry to the grandstand.

The unveiling of this scheme, together with later variations upon it, led us to our present policy on refunds, which is not to make any, except in cases of obvious sudden illness or accident.

After the races, there were still the payoffs to be dealt with. The purse was based on a percentage of the gross receipts after taxes, and was distributed according to a schedule prepared by the American Racing Drivers Club. In addition to this amount, there was always an additional disbursement of $200 to $400 in goodwill offerings, appearance money to drivers who had come but because of some difficulties mechanical or otherwise were unable to race, and crying money to owners whose cars were damaged or to those who, due to some accident, failed to qualify. The boys expected to receive some token payment for trying and, if it wasn't forthcoming, the field might be too small the following week to present the races as advertised.

It was midnight before the last driver's envelope had been prepared and delivered to him by the steward. Then we all adjourned to the better-equipped president's office for cheese and crackers and the inevitable discussion of the evening's card. It was a lot of work, but a lot of fun, too.

For the bugs, who kept multiplying nightly that fall, John kept setting up new bleacher sections at either end of the grandstand until we were able to accommodate a record crowd of over 10,000. The races ran into the first week of November, so late people began to call them "overcoat meets," but nobody seemed to mind the cold. It was income from these and the next season's races that enabled us to pay for so much restoration of the grounds and buildings in preparation for the 1946 Fair.

With the midget races still in full swing, John began carrying on what I thought was a mild flirtation with speedboat racing. We made excursions to Lake Onota near Pittsfield, Massachusetts, and to Agawam on the banks of the Connecticut River. We wandered as far as eastern Pennsylvania, where John had heard of a small round pond serving as a course for the smaller classes of these craft. We spent one whole summer of sunny Saturday afternoons and Sundays in this agreeable dalliance before I became aware of any special significance in the trips.

Of course I should have been suspicious, but on what basis? Except for three lily ponds in the New England village, all the water on the fairgrounds lay between the brushy hummocks of the swamp. That some of it might be transferred by pumping to the infield of the half-mile horse track was beyond the scope of my imagination at the time, but that is what happened

The midgets race on the fifth-mile track, early 1950s. Note the waterway surrounding the track and the bridge crossing to the infield.

The land was low there except for some rising ground in the middle. A couple of weeks of excavating made it considerably lower. As soon as the dirt had been trucked off to fill depressions in the parking areas, two streams flowing through four-inch pipes began the flooding process. Slowly, the ring of water rose and broadened until the whole expanse was flooded except for the center area where the midget track lay. At the north end of the grandstand, where the channel was somewhat narrowed and confined between side walls of cement blocks, a stout wooden bridge was built by which to reach the newly-formed island. The average depth of water was three feet, more than enough for the tiny speed boats, whose avowed draught was eight inches, but that skipped about the surface like water-skaters and could navigate quite airily wherever the grass was wet.

Outboard hydroplane racing was the form of this sport with the largest number of participants, probably because it involved the minimum outlay of cash for the newcomer. Its classes were based on weight of hull, exclusive of engine, plus weight of contestant. Thus in the M or midget class, with a minimum hull weight of 75 lbs, the overall weight for boat and driver was 200 lbs. In classes A and B, the hull weight was 100 lbs with overall weights of 250 and 265 lbs respectively. There were also C and F class boats, each with a little heavier hull than the preceding class to absorb the weight of a heavier driver.

Contestants were drawn largely from among hydroplane owners, although we had a class for inboards. The boats, arriving on trucks and trailers, were unloaded on the far side of the waterway, where a dock had been built for pit use. The racing was conducted in midget auto style with the fastest qualifying boats started in the rear and with auto-type flag signals used. Drivers wore crash helmets, lifejackets and rubber knee pads, the latter to cushion the pounding of the hull in competition. Prize money, distributed by the club, was based upon a percentage of the gate receipts.

Wednesday evenings were selected for the boats as least likely to conflict with the Saturday night auto races, and the four meets held during the summer of '48 were encouragingly well-attended and received. Under the lights with the spray flying as drivers crowded each other for a pole position, the racing was exciting enough to please anybody.

As a spectator sport, however, it did reveal certain weaknesses. The outboard motors were not as immediately responsive to the pull of a string as automobiles are to a self-starter and the lining-up of these volatile waterbugs often took too long. Accidents in the narrow homestretch, while unlikely to result in serious bodily injury, were destructive to the light wooden craft, which frequently required more costly repairs than their owners could regularly afford.

That fall, John started boat-building experiments designed to augment and strengthen the inboard class. One end of our grinding shop,

The boat races at the "Aquaway," 1948.
Bottom photo *Danbury Museum and Historical Society*

where work had slackened since the war, was already given over to a new process in plastic modeling. There Harold Kohler, John's cousin, was successfully creating dozens of the larger-than-life ornamental figures that decorate the fairgrounds and its buildings. Taking what appeared to be the ideal hull as a pattern, Harold reproduced it in light, strong, splinter-proof plastic. Into his total winter's output of eight crafts, Charles Gereg and Everett Hislop, our super-mechanics, installed Crosley motors with self-starters. By June, they were ready for action.

The races were, indeed, improved by the new inboards as John had expected. Starts took less time and the various events went off on schedule. Contestants, formerly cautious from experience with the hazards of the course, but now freed from anxiety over repair bills, developed some dashing techniques in driving and a latent flair for showmanship. The fans began to have favorite drivers for whom they watched and rooted. For a while, we had high hopes.

These were not, however, like the early days of midget racing, when stands and bleachers were jammed to capacity. After a few weeks of auto racing on Saturday nights and boat racing on Wednesdays, it became increasingly clear that both shows were suffering from a division of patronage. The racing public either would not or could not afford to support two shows a week, and the more firmly established auto races had the inside track.

The starter at the boat races, 1948.

Since from an expense standpoint it is better to fill the grandstand once a week than to have it only two-thirds occupied for each of two separate shows, the boat races were called off. However, we still have our stable of boats as souvenirs.[51]

51 The fleet of 18 boats was sold to private collectors in the late 1970s.

13

THREE TO MAKE READY

The war's end in 1945 brought about my gradual release from the New Milford business. After that winter, John took the calculated risk of spoiling me with a vacation. For a few weeks that spring I visited, shopped and repaired my depleted wardrobe as a proper lady should.

Then Mrs. Keating, who had been the prop and mainstay of our home life, made up her mind that she, too, needed a rest. She denies that having me around the house so much was a factor in her decision, but it sure did look funny!

When at last I had my chance to experiment with housekeeping, at which I had always confidentially expected to excel, I quickly saw through those Russian women who take so readily to laying bricks. In a job where my reputed flexibility should have stood me in good stead, I went around all day in circles and my feet, unaccustomed as they were, hurt unmercifully. What caused me further chagrin was that John, realizing far less than I had about the complexities of housework, behaved as if he thought I was still out to grass.

As he had been doing for three summers now, he daily set out for the fairgrounds at 7:30 a.m. in his battered Ford coupe. The Ford and a white baseball cap are John's distinctive markings and have always saved time and mileage, since a glimpse of one or the other in the distance is positive identification. Usually an hour or so after he left, just as I was deciding to whip up something for dessert, he would call me for the first time. If the following isn't a fair sample of what took place regularly, it is a very close approximation.

"When you come over with my lunch, would you stop and pick up a few things at the hardware store? Got a pad? We need six galvanized water pails, ten cheap hammers, a dozen red dustpans, two boxes of quarter-inch stove bolts and a few hacksaw blades. Got it? Then go to the D.&B. Lumber Company. They'll have a bundle of shingles ready to put in the back of the car. Come by the way of Franklin Street and see if Mr. Henebry has the small castings ready."

I scribbled away.

"What time will you be over?"

"Well, I'm a little busy right now, but by noon sharp." Three stops, with one on White Street where the parking was tough. That scuttled lemon meringue pie, but there was always a box of Jello and some fruit to cut up in it.

In no time it seemed, the phone rang again.

"Going to need a few more things downtown. Harold wants five pounds of ten-penny nails, six boxes of one-inch staples and a 250-foot coil of manila rope. Seaman and Lynch have the rope. Bill wants five gallons of outside white and a gallon of turpentine. When are you going to start?"

"I was going to leave at eleven, but I'll have to go everywhere for nails. I'd better get going."

Nails were my bugaboo that post-war summer. They were scarcer than two-headed calves, and I had to buy a pound here and a pound there. The whole country had gone nail-crazy.

"When you get here," John continued, "stop by the tool house. I'll have someone help you unload."

"That's just dandy," I assured him, "if you're certain you can spare the help. And how would you like some canned raspberries for dinner?"

Just as I was picking up the brown paper bag of lunch, Mr. Jarvis telephoned from the White Street office.

"John just called me. He forgot to tell you to stop here for a roll of wire. It will be down back at the yellow door. And please bring his blue suit, the brown shoes in his closet, a clean shirt, a tie and socks, and a few handkerchiefs."

"Is he leaving town? Why didn't he say so?"

"No, but there's a meeting of the Antique Dealers' Association over there this afternoon and he doesn't have time to come home and get dressed up."

"Did he mention a carnation for his buttonhole?"

Irv giggled slightly. He always treats my most biting sarcasm as mere roguish jest and frequently forgets to quote me where it will do the most good.

C. Irving Jarvis, John Leahy's right-hand man.

The nearer White Street draws toward Main, the more clogged it gets with trucks double- and triple-parked trying to make deliveries to stores that have no rear entrances because they back up to the Still River. Both lanes of ordinary traffic are often stopped completely and then a great ruckus of horn-blowing arises and continues until some blasé truck driver finishes unloading, lifts his hand truck aboard, slams shut his tailgate and moves on.

In this area of frustration are situated most of our paint, plumbing and hardware establishments and I had early developed the eye of a hungry hawk for a parking place about to be vacated. Once parked and inside a store, I became an object of interest to the personnel, somewhat on account of the quantity of my unladylike purchases, but more because of the hearsay and speculation about town as to what John's actual plans were for the fairgrounds.

Clerks, who obligingly carried to my car the hammers and nails, brushes and paint, rakes and saws, had seized the opportunity to verify or disprove the rumors.

"They say John's going to build oil tanks along the railroad siding over there," one youth suggested from behind an armful of pails and brooms.

"They do?"

"Yep," that's what a salesman told Emil yesterday, "but this morning another one from Bridgeport said John's selling fifty acres to Cities Service and they're going to locate a bulk storage plant right opposite the main entrance."

A pause here for my comment.

"You don't say?" I exclaimed. "He must have read that in the Bridgeport Herald."

He stopped short and set down his burden.

"Was it in the Danbury section?" he demanded. "Funny, I didn't see it."

"Not that I know of," I said. "I just meant I hadn't heard of that myself."

"Well, I didn't take any stock in that Cities Service yarn," he admitted. "Why would they want to build big oil tanks in Danbury. They'd want to be on the shore, wouldn't they?"

"I should think so."

"What sounded more likely to me was that the state is taking over the land to add to the airport. That wouldn't surprise me. For training interceptors, you know."

"The state is going to train interceptors?" This was too much and I let the wet blanket descend. "Is New York going to attack Connecticut? Is Dewey hatching up a war?"

Whereupon, since he had carried all my heavy purchases, I felt a little mean.

"Sorry, Mike, but I haven't heard that one either and I'm sure John hasn't. It's just as well, in a way. It would only worry us."

As I drove up the main midway at noon, John met me to extract from my collection what he needed immediately. Then we drove around to drop off the remainder of the load.

The better to control the supplies, certain small sheds had been assigned to hold carpentry, electrical and plumbing equipment. There was a paint shop, too, and a general tool house. Even then, it was impossible to keep track of the various small items that were continually used up, worn out, misplaced or that just walked off, so that supply never caught up with demand.

Any qualms that John might have felt about the disruption of my morning, he concealed completely.

"Good!" he exclaimed. "Fine," as I produced four little packages of nails, each representing the good will of a different hardware store. Then he drifted lightly to a more innocuous topic.

"That Schoen boy I hired last week is O.K. No trouble anytime to find him."

"Why should there be?"

"Oh, some of them run and hide when they see me coming. When I see a feller run toward me, I'm pretty sure I can depend on him."

"And the opposite sex?" I asked him, with mounting conviction that I stood on the brink of discovery.

"I like 'em that way too," he admitted with a grin.

Thus having fathomed the secret of my charm for John, I felt bound to remain on call like a fireman, even though my mornings were altogether lost to homemaking.

The more I catered to it, the more the Fair became my baby as well as John's, and I wound up by making a daily tour of the acreage.

By midsummer 1946, the major work of restoring the buildings was finished.

The grounds were enclosed by a new eight-foot woven wire fence that looked strong and businesslike, as befits fencing that guards the gate receipts.

Asphalt had been laid on the principal midways to do away with the clouds of dust that formerly had coated merchandise and customers in dry weather and made a horrid quagmire when it rained.

The canvas of the Big Top had been mended and flame-proofed.

A picnic grounds had been laid out on the smooth green area to the rear of the Sportsman's Building ready to be strewn with stout trestle tables[52] and benches, most of which had been manufactured from odds and ends of used lumber. Two coats of blue paint concealed their shabby past.

Bleachers would later be set up on three sides of the level plot where the ox drawing customarily took place.

There were also new park benches waiting to be arranged on the plaza and scattered along the walks. John and I had by that time plodded enough miles at enough fairs to realize that nothing is more disheartening than aching feet, and a patron who goes home disheartened may vent his distress in unjust criticism.

Inside the gates, the grass had been kept clipped all summer, and on the parking fields outside it had been mown. Just before the Fair opened it would be cut again.

So much for physical preparations.

The amount to be spent on advertising was an important question on which John, as usual, had gathered a grist of opinion and then used his own judgment.

Mr. Rundle's advice was to keep the budget low.

"Just let it be known the Fair is going to open up. They'll come," he said. "They always did."

This counsel tickled John. He quoted it, but he was inwardly skeptical.

"After you've been away four years, only your mother wants to see you again," was his opinion. "We'll have to spend money on the first Fair."

Spreading the word locally was easy, but to be successful the Fair must draw from a radius of at least sixty miles, which included Hartford, New Haven, Bridgeport, Waterbury, the Norwalks, Stamford and the Hudson River towns from Yonkers to Poughkeepsie and New York City.

To cover these communities, our publicity was entrusted to a professional, Walter Bull of the *Illustrated Speedway News*, who had done a good job for us with the midget races.

For newspaper, radio and billboard advertising, the budget was set at $12,000, not much perhaps to the manufacturer of a filtered cigarette, but a lot to us.

Ticket selling and taking had formerly been done by local applicants for the jobs, who had always come as they were and done the best they could at it. Some had done very well indeed.

John had, however, chosen Pinkerton men for this work, both on account of their neat uniforms, which he had noticed at other fairs, and because they were strangers in town, and so under no social obligation to admit friends who proffered matchbooks in lieu of tickets. The Pinkertons were also put in charge of parking.

---

52 A table consisting of a board or boards laid on a framework of a horizontal beam supported by two pairs of legs, often sawhorses.

Mr. Harold Crowe, an officer in the Danbury National Bank, had been persuaded to take part of his vacation during Fair Week in order to "assist" me in the treasurer's office, a phrasing for our relations that, as we both well knew, was certainly putting the cart before the horse. Not only was he a master of addition and subtraction, but he had a way with the customers that sent them away delighted, whether he cashed their checks or not. That, I believe, is the pinnacle of finesse. He proved himself a friend indeed by his extraordinary tact and patience, never having yet lifted a hand or foot to me, as he would have been justified in doing considering my mistakes. I could gladly have lifted a foot to myself on many occasions.

The booths in the Big Top were rented and only some of the less desirable space outside remained unsold. Even the least fervent of our sidewalk superintendents had to admit the Fair would probably open, although they stuck to it that it wouldn't be what it used to be. Of course, they were partly right about that. Everything changes with the march of time, including prices. As the cost of advertising would be higher than in former years, so would other running expenses. Premiums, wages and grandstand entertainment all would reflect the upward trend.

The break-even figure had kept increasing every time we counted our chickens, which we had begun to do when the war ended a year ago. As Fair Week drew nearer, it reached a point that would have shocked the Old Guard and which caused John to remark that he was born thirty years too late. He should have been a Mississippi River gambler.

It would take a good big gate to deal with such large bills. We were especially interested in the first day's attendance as a barometer of the Fair's prospects.

The records of the past five years, 1937 through 1941 had shown an average opening day attendance of 24,000, but those records were approximations we were warned, rather than a close count of paid admissions. They represented the total number of persons on the grounds estimated from the count of tickets taken at the gates, the daily passes taken, the tickets torn off weekly passes, officials who came in on their badges, uniformed employees like police, and organizations like bands, which came in a body without passes.

In Mr. Rundle's time, when the newspaper reporters cornered him for an attendance figure late in the day before the small army of lady helpers upstairs were through sorting and counting the tickets one by one, he would squint over his glasses out the window and then at the big clock on the opposite wall and answer right off the bat, "19,428."

His estimates were surprisingly accurate as a result of long experience with the eyeball method, and what was good enough for him had always been good enough for the newspaper. His chickens were pretty well hatched by 4 p.m., but the correlation between them and the cash receipts was not so quickly arrived at, if indeed it ever was.

After much winnowing of facts from the chaff of hearsay and shaky statistics, subtracting a few thousand here for passes and adding a few thousand there for full employment, John emerged from all this guesswork with a wishful estimate of a 25,000 paid attendance for our opener.

Everything now depended on the weather.

14

OPENING DAY

A little man in the back of my head who hates an alarm clock called me on the dot of 6 a.m. so that I could switch the thing off before it rang. Then he pestered me till I opened my eyes.

No light filtered into the room around the edges of the window shades.

By slow degrees, I sat up and peered over the footboard toward the field of observation provided by two inches of open window. The result was unsatisfactory.

Then suddenly I remembered what day it was and why the weather made a difference.

I cleared the bed with one brave bound and rolled both the shades way up.

The morning was thick with heavy fog and the trees dripped softly.

"And 25,000 people are going to turn over and go back to sleep," I murmured in distress.

"Talking to yourself?" called John cheerfully from the vicinity of the bathroom.

"It's raining," I mourned.

"It rained in the night, but what you hear now is just water dropping off the leaves. You'd better start getting dressed," he advised. "You're not among that 25,000."

Setting about dubiously to follow instructions I turned on the radio. The weatherman had been undecided the night before.

"Rain," ran his prediction, "ending tonight or tomorrow."

Now he was slightly more optimistic, but "gradual clearing in the forenoon" was the best he could do.

We had the lights on in the dining room while we gobbled our coffee and toast. In fact, they were on all over the house and I had to go to the front door for a look at the outside atmosphere, which close-up seemed gradually to be undergoing a change for the better. From soupy, it was turning milky.

As we backed out of the garage, our next-door neighbor hailed us. "This fog is going to burn off," she yelled. "Good luck!"

We waved gratefully and hoped she was right.

It took us fifteen minutes or less to reach the fairgrounds. I shall never forget how in that short interval the mist completely dissolved so that, as we entered the main gate, the day was beginning to turn golden with sunshine, tangy with the breath of fall.

In my treasurer's office on the second floor of the Administration Building, there were rolls of tickets and bags of change waiting in metal boxes for the twelve sellers. Their number had been based on the hopefully estimated attendance of 25,000, which would have pleased and gratified all concerned. Mr. McKenney, in charge of the Pinkertons, had thought ten men could take care of the selling, but to be on the safe side he had brought twelve.

They trooped in at 8 a.m., fresh and perky looking as Air Corps Cadets to whom they bore a notable resemblance in their blue-gray uniforms.

Before eight o'clock, people were waiting outside the gates and as soon as the ticket windows opened, the lines through the turnstiles became steady processions that never faltered.

By 10 a.m., the sun was getting high and the sellers were fighting a losing battle against lengthening lines that kept moving, but not fast enough. By 11 a.m., their neat collars were wilting and they were calling for reinforcements, which arrived only as Mr. McKenney and his first mate began to sell from bags, and sympathetic concessionaires rushed hot dogs and coffee from nearby stands. Not one of them had time-off for lunch that day. They said it was like trying to bail out a leaky boat.

The populations of eastern New York and western Connecticut seemed possessed by the idea that a fair was just what they had been missing most ever since Pearl Harbor.

On either side of the new avenue running from the Arena Gate to the Administration Building was the carnival area. There Oscar Buck and Ross Manning had their shows, games and rides.

Up this midway set the main current of the multitude with rivulets streaming out from it toward the pony rides, the merry-go-round or the Silver Streak. Every sideshow along the way subtracted from the crowd without visibly diminishing its volume. Through the medley of farmers and fakers, of townspeople and the station wagon set, strolled hawkers with huge bunches of balloons and cowboys in ten-gallon hats.

"Hurry! Hurry! Hurry!" a barker urged all within the radius of his microphone, and that was a good many. "The show is ready to go on. See the little lady handle the poisonous reptiles. She has rattlers. She has copperheads. Come in and see her charm the cobra, the deadly snake of India. These venomous reptiles coil about her limbs, her waist, her neck. What nerves she has. What curves she has! Hurry! Hurry! Hurry!"

"Just about to feed the wild man," shouted a competitor who was brandishing a piece of raw meat on a pitch fork. "Hurry, hurry! See the wild man of Borneo. He eats 'em alive and he eats 'em raw."

"All hot! Get 'em while they're hot!" came a volley from another

The main plaza with the Administration Building and Fair entrance, 1941. *Frank Baisley*

The midway offered a variety of options to the fairgoer, 1930s. *Frank Baisley*

Father and son dressed for the Fair, 1933. *Frank Baisley*

quarter. "Some don't like 'em raw. We got 'em cooked, franks or sausage. Get 'em while they're hot!"

On a raised platform at the entrance to the Big Top, John had mounted an antique mechanical cow as a sort of Fair motif. She looked like a slightly oversized Brown Swiss and was said by the antique dealer who sold her to have been made in Switzerland for the Harris Milk Company as an exhibit at the Chicago World's Fair in 1893. She contained a 25-gallon tank from which she gave milk for the people who had visited the Chicago exhibit. Lacking faith in her mechanism, John omitted to fill her tank, but instead fastened a rope to the cowbell that hung from her collar and tacked a sign to the platform. "O.K. for little boys to ring bell."

Farrow as she was, she enjoyed a remarkable popularity. Everybody with a camera snapped a companion and was in turn snapped milking the cow.

The cowbell rang constantly.

The new Administration Building and 4-pole Big Top, built after the 1941 fire. *Frank Baisley*

The Big Top has always been the heart of the Fair. There beneath its soaring canopy are the most valuable exhibits and concession spaces. Sooner or later, every visitor enters the high-arched passageway that bisects the Administration Building.

On the long tables facing the entrance were the dahlias and chrysanthemums, those hardy survivors of all but the blackest frosts. Next came the Grange displays, each topped with its own title and theme of good husbandry in purple letters on a white ground. To right and left, presided over by Howard Shepard, were sights to justify the most unlikely of all seed catalog illustrations. For here in brilliant array was the aristocracy of the fruit and vegetable kingdom, the apples and quinces, the grapes and the plums, the pumpkins, the squashes and the lacquered gourds, the orange carrots and the long taper fingers of green beans with the purple of eggplant and the gold of corn.

Nearby, under the direction of Mrs. Pauline Tilson, was the women's department, where, suspended from on high, were crocheted tablecloths and bedspreads, patchwork quilts and hand-hooked rugs. Beneath, on wide tables and in showcases, were exhibits of needlework both plain and fancy, including many a masterpiece of knitting and crochet.

Here, drawn up for inspection and carefully labeled, stood myriads of glass jars, quarts and pints crammed with baby beets, green and yellow snap beans, and whole ears of sweet corn. There was a sprinkling of canned fruit too, strawberries and raspberries, peaches, cherries and plums. However, sugar was so scarce in the autumn of 1946 that jams, jellies and preserves were conspicuous by their absence. There were, as a result of the sugar shortage, enough ribbons to go around to all the bakers of cakes and pies. But prize bread and rolls, brown and tender crusted, testified that the dearth of sweets was no fault of local housewives.

This Big Top display from the 1930s included gourds in the foreground and leafy vegetables in the back. *Frank Baisley*

The circular building that forms the sides of the Big Top has an inner ring of booths facing the center area and an outer ring forming an arcade where the booths face each other across a walkway about 12 feet in width. These spaces are rented for purposes of mercantile display or advertising.

Not to be outdone by the vegetables, the floral displays were just as extravagant, 1949.

Here numerous local tradesmen set up shop for the sale of anything from a Danbury hat to a Castro Convertible, but a great deal of merchandise travels to the Fair from a considerable distance. There are maple products from Vermont, shell jewelry from Florida and countless items from the states in between. There are the *Little Golden Books* from Brentano's and the big Hammond organs from John Wanamaker's to remind Danburians that the metropolitan area is fast extending its range toward this neck of the woods.

The Republican Women's Club has a booth on one side of the front entrance, while the Democrats are installed a few spaces down on the other side.

Even as the sightseers marveled at the milk-fed pumpkins, took stock of the bedspreads or succumbed to the wiles of a pancake salesman whose patter was reinforced by samples, they were regaled by band music. Morning and afternoon concerts had been the custom since anybody could remember, the idea being no doubt that a certain gaiety is thereby engendered, the spirit of the occasion and all that. Personally, I prefer to take my music in a more relaxing environment. I avoid the supermarket that distracts me with "Memories" while I try to remember what I came for, but that is neither here nor there.

John rather likes band concerts—it must be the German in him—but that year he varied the routine by introducing square dancing on the same center stage used by the band.

Young Al Brundage of Danbury's own King Street had assembled a country dance orchestra comprised of piano, banjo, fiddle and drums, and at the age of 26 had become champion caller for the State of Connecticut. In blue denim overalls and polka-dotted shirts, the young musicians evoked as much hilarity by their verve and antics as by their music. Tude Tanguay, a three-year veteran of the Army Transport Service, did some acrobatic accompaniments that included fiddling lying on his back, on one elbow and finally standing on his head, each attitude proving progressively more uproarious to his convulsed audience.

This innovation was so merrily appropriate and such a hit that Al and his supporting cast have become a yearly attraction at the Fair. He has since built The Village Barn in Stepney for his square dances and has gone on to widespread fame via television appearances.

The western exits of the Big Top were the busiest since they faced the new Sportsman's Building with its wildlife exhibits and the Poultry Show, where the Snows, Leland and Jesse, superintended 2,000 cacklers of all known breeds.

A new walk, christened the Garden Path, ran between these buildings through the new picnic area in the direction of Machinery Hall. Nearby were the concrete cattle barns, the pony barns, the calf nursery and the sheds for swine, sheep and goats.

Those who finished their tour of the Big Top on its eastern side emerged upon Church Row, so-named in former years when denominational dinners were the order of the day.

These restaurants, now privately operated and closely supervised by the State of Connecticut with a finger in every dishpan to test the amount and temperature of the water used, included Swedish, French and Italian cuisines as well as conventional Connecticut cookery. They sent up such an international blending of aromas as to suggest immediate safeguards against the onrush of starvation.

The two slices of homemade bread and butter stuffed with warm turkey meat with which at midday, I fortified my strength and spirits, came from one of these places and still dwell so deeply in my hungry senses that no between-meals snack has ever seemed quite as good.

As the rate of individual progress from one exhibit to another gradually ceased to be a matter of intention and became geared to the motion of the crowd, it was evident that our 25,000 hoped-for patrons had not let us down.

Long before noon, traffic on the main roads into Danbury was slowed down and, by early afternoon, the state police were turning back cars on the Norwalk road to the south. Through-traffic from the north was shunted to Long Hill where lay the nearest entrance to the Merritt Parkway. So great was the congestion that cars were pulled up on the shoulders of jammed highways or left parked in the city streets while their occupants continued the journey to the fairgrounds on foot. In blaring iteration from loudspeakers on police cars was issued the instruction, "There is no more parking space on the fairgrounds. Keep going straight on the main highway."

As a matter of fact, there were some parking spaces on the grounds at that time since many patrons who had come early were beginning to start for home by 2 or 3 p.m. The truth, pure and simple, is that there are not enough roads in and out of Danbury to take care of normal weekend traffic. What we have are too narrow. No wonder the harassed state police were content simply to keep the procession moving!

By 2 p.m., food supplies were running low and concessionaires dispatched couriers into town for additional amounts, which on Sunday were hard to come by.

By 4 p.m., even the big restaurant under the grandstand run by Walt and Russ, the most experienced and optimistic of purveyors, was sold out. Finally there wasn't a frankfurter or a hamburger left on the fairgrounds and very little else to eat or drink.

Afterward, I learned from John that the old dug well near the farmhouse had early given its all, and the new artesian well near Machinery Hall, which supplied that end of the grounds, had also failed. There was water left in the swamp, of course, but of the three big pumps that worked all day to supply water to the batteries of toilets at the north end and under the grandstand, only one had continued to deliver. The others having burned out their motors and quit, creating an acute shortage and a frantic rush job of finding and installing replacements. It was a good thing we weren't running a night fair.

We stop selling tickets at 5 p.m. and the Big Top closes at six because October days are chilly after the sun goes down. By that time most people are weary enough to want to go home, although a small crowd of latecomers linger about the midways for an hour or two later on the warmer evenings.

From closing time that night until after 11 p.m., the treasurer's office came in for its share in the general delirium, but by then, while I was toiling upward in the night, John was as happy as a kid on a swing with a popsicle. He had been right. People still enjoyed a country fair.

At last, the differences between my dubious bookkeeping and Mr. Crowe's cash receipts having been tentatively reconciled, the fact was established that our paid attendance for the day had been close to 42,000, which constituted an all-time record.

15

LIVE AND LEARN

That night a cold northeaster swept down from Labrador or wherever northeasters breed with a raw wind and sheets of driving rain. Thankful that its arrival had been postponed until Monday, which at best is a slow day, John set diligently about making repairs and better provisions for contingencies no longer unforeseen.

Sunday had been so mild that there was no reason for getting up steam in the huge boiler that furnished heat to the main buildings. The boiler itself, housed along with the water pumps in a neat shed at the north end, was an aged refugee from some defunct hat shop, but had, of course, been tested under pressure. The pipes that ran from it, however, lay underground, so their condition was not thoroughly ascertained until that Monday morning when, back in service after their long retirement, they ambitiously attempted to warm the outside world as well as the inside by emitting at intervals from the ground between the pump house and the offices periodic jets of steam. You'd have thought it was Yellowstone Park, and I suggested that we bill the largest as "Old Faithful" and let matters take their course. But John had the worst leaks patched up. Then, after renovating the pumps, the toilets and some broken water pipes into which cars had backed, he turned his attention to making some baby carriage gates.

John Leahy loved his Fair, and was not above mopping up when the need arose.

On the previous day, progress at the turnstiles had been hampered by an influx of baby carriages and strollers, since no aisle was quite wide enough for their entrance. Whoever planned the gates must have done so unaware that, after the Japanese surrender, the Spirit of '46 was best to be symbolized by the baby carriage. The result was that one man had been kept busy the entire day opening and closing a drive-in gate or boosting these vehicles over the turnstiles.

Inside the fairgrounds, the baby carriage has its advantages, since it affords a convenient receptacle not only for one or two small fry with bottles and diapers, but also for lunches, cameras, thermos bottles, cushions and other picnic paraphernalia. On the way home, it will accommodate prizes and purchases.

Energetic young fathers and mothers seemed to think nothing of bringing an infant in carriage, a toddler in stroller, and older progeny wholly or partially self-propelled. Such hardihood deserved recognition and gates were hurriedly rearranged with one wide aisle labeled "Baby Carriages."

The weather, instead of improving, which would have expedited the repair jobs, grew wetter and worse all day. Just the same, a few dauntless spirits paid their admission and resolutely slopped about from building to building.

At noon, I scuttled across the plaza to Church Row for a sandwich and coffee. As I waited, shivering on my stool among other glum-faced coffee customers, along came a small sailor and perched there beside me.

I must have reminded him of the little match girl, for he resolved to warm me up.

"What you need, lady, is a shot!" he exclaimed feelingly and before I realized it he was lacing my coffee from a pocket flask.

"That's enough," I weakly protested through chattering teeth, being too polite to object further.

"Don't be stuck up," he admonished me, subjecting his own cup to a similar dosage.

That I denied.

"But my boss might not like it," I added, reflecting with some amusement on the reaction of my banker colleagues to a whiskey breath.

"Always carry a life preserver," continued my gay young mariner. "Who's your boss? Mine's Roosevelt."

"Mine's Mr. Crowe," I replied. "You may not know him, but I feel sure he's quite a bit fussier than Roosevelt."

I never expect to learn the identity of this companionable lad, but I wish he could know in his old age how much my dark day was brightened by that brief encounter.

Tuesday was likewise cold and cloudy, but the rain had ceased to fall.

That was the day the leader of our high-priced band discovered that his presence was urgently needed elsewhere and came to the office wildly flourishing a telegram, which, despite an earnest effort, John did not get to read. Family difficulties, he claimed, had arisen in Boston. Joe must go.

"My brother, he conducts too. I go now, come back Thursday."

"Let your brother go to Boston," argued John. "This is *Joe* Muscanto's band. Nobody knows your brother. You stay here, let him go."

"My brother conducts fine. Lotta times I go, he conducts. You'll like my brother. I go today, be back Thursday."

John is not one to give up without a struggle with the result that the spluttering flow of Joe's language rapidly became inadequate for the release of his emotion. Just before he reached the bursting point, John saw the light. Joe wasn't asking him, he was telling him. The fine print of his contract, it seemed, did not call for his actual presence every day, but only for that of his name band.

So Joe went and his brother conducted until Thursday afternoon when he appeared before John in the same state of excitement that had heralded Joe's departure. He too clutched a telegram and must go to Boston.

"The trombonist, he will conduct. He conducts alla time. You'll like him," asserted Joe's brother. "Joe and I, we both come back Saturday for the weekend." Then he, like Joe, decamped. Since the band played on, apparently unaffected by the presence or absence of conductors, John reluctantly decided to like the trombonist, which was all he could do for the time being.

On Wednesday morning, the sun at last broke through in welcome, honoring our first Governor's Day.

After their lunch downtown with the Rotary Club, Governor and Mrs.

Baldwin[53] came to the Fair riding at the head of a cavalcade of motor cars escorted by state police whose motorcycle sirens rambunctiously cleared the way and announced their advent. In their wake followed the town and city officials, who, regardless of party affiliation, yearly bury the hatchet and unite upon this occasion in loyalty to the Nutmeg State and in honest pride of their own tolerance.

Through the main gate, this entourage rolled with 10,000 pairs of eyes upon it and a like number of hands raised to applaud. Up to the Administration Building rode the Governor and his party, where John waited beaming like a junior Grover Whelan[54] to enact his duties as host. Into the Big Top they passed, where photographers lay in wait to record their acceptance of flowers and fruit. When they were enthroned at last in a center box on the grandstand among their minions, Joe's band struck up a creditable version of "Hail to the Chief."

The new Administration Building with the Big Top poles in the background, 1941. *Frank Baisley*

The Governor was then introduced and spoke briefly. A pretty good show in itself, smacking somewhat of the medieval, I reflected, speculating light-mindedly to myself on what a few heralds, say half a dozen, with trumpets would do for the production. If handsome Lt. Sullivan, for instance, could be persuaded to lay aside his blowtorch,[55] we could easily scare up five patrolmen with the proper qualifications.

Without further trimmings, however, the ritual seemed to delight the audience as much as the more professional entertainment that followed. Just as the governor had received a royal welcome, so he left in a blaze of glory. My sympathy and admiration were all for Mrs. Baldwin, who, throughout a routine that long ago must have become tedious, remained gracious, kindly and attentive as a good wife has to be.

Our spectator attractions had been carefully planned under Mr. McLean's experienced guidance. Automobile thrill shows were known to be popular here and two were on the schedule.

---

53 Raymond E. Baldwin (1893-1986) served as Connecticut's governor from 1939-41 and 1943-46.

54 Grover Whelan was New York City's "official greeter," chief of the mayor's Commission for Protocol for 34 years.

55 The concessionaires were required to use flame-retardant materials for their exhibits. State Police would patrol the grounds with a blowtorch, testing the exhibits to make sure they wouldn't catch on fire.

Earl "Lucky" Teter, founded the Hell Drivers in 1934, creating a massively popular show filled with precision driving and deliberate crashes, 1938. *Frank Baisley*

Lucky Teter in T-bone crash, 1938. *Frank Baisley*

The thrill show is a headlong display of recklessness in which automobiles do broad jumps, roll over, crash through burning barriers and collide with each other.

Danbury had always filled the stand to see Lucky Teter, who up until 1942 was the foremost exponent of car crashing. Lucky, who started his career as a test driver, had assembled a group of young daredevils and trained them in stunt driving. Although each was a master of some hair-raising specialty, their leader always reserved the most dangerous feats for himself.

His show played Danbury for several years, including 1941, the last year the Fair operated before the war. Up until that day in Indianapolis when, attempting to jump his car over a bus parked lengthwise, Lucky Teter unluckily crashed into a ramp and lost his life, he was unequalled in skill and showmanship.

Strange as it may seem, he had a number of capable successors, among them Jack Kochman and Joie Chitwood, who presented their shows on Wednesday and Friday that season, respectively.

Every afternoon, there was quarter horse racing, an innovation in our region. Western horses, employed in cattle roundups are trained for quick bursts of speed. The best of them are matched at rodeo time on quarter-mile courses, from which they receive their name. This racing, under the direction of C. J. Walter of the Cinnabar Ranch, featured western saddles and equipment.

Preceding and interspersed with these featured exhibitions and presented within the race track oval were six specialty acts, introduced by Georgiana Dieter, a lovely and competent mistress of ceremonies.

There were two animal acts, Woodford's Dachshunds and Jack Andrews' educated Brahma bull, Henry. There were "The Briants," sensational acrobats, and "The Four Macks," really remarkable figure skaters. There was Ben Mouton, whose breathtaking calisthenics took place on a swaying pole 160 feet in the air, and there were the Berosinis, billed as the royal family of the high wire.

We shall long remember the Berosinis, and not for their technique, which was faultless as advertised, but for an oversight on the part of their agent, who neglected until the last minute to mention the two indispensable props for their act.

John W. Leahy enjoys sitting on Henry, the educated Brahma bull, owned by veteran rodeo executive, Jack Andrews, 1946. *John J. Heyde*

At 4:30 p.m. on the Saturday afternoon before our opening day, John received a telegram advising him that two large hooks must be strongly fastened to cables secured in the ground a certain distance apart. It was to these hooks that the Berosinis' high wire would be guyed.

Digging the two large holes and burying the crossed railroad ties that served as cable anchors represented several hours' work. Everybody was tired. It would soon be dark and John was personally beset with countless problems that, at best, would keep him going half the night. This was more than he felt able to tackle, and so he didn't tackle it.

In the morning, the Berosinis arrived ready to attach their wire for the afternoon performance. Where were their hooks, they plaintively inquired of all and sundry until they caught up with John, who was busily directing the transfer of a chimpanzee's watery playpen to higher ground. He tried to explain, but was quickly silenced by a barrage of reproach that struck him as excessive in proportion to his crime. At the moment, his attention was somewhat divided by the claims of Joe, the chimp, whose ability to express himself succinctly without speech possibly inclined John in his favor.

It is hard to continue a one-sided argument, and these fine-muscled athletes changed their mode of approach. They produced and cited their contract. Had John read it? He had not. He considers reading such contracts as a waste of time. On the other hand, chimpanzees fascinate him.

The Berosinis pointed out that while they furnished themselves and the wire, his share was to furnish the "appurtenances thereto" and pay for the act "regardless."

When John replied that he realized his financial responsibility and was negotiating with the Connecticut Employment Bureau for hole diggers, all recrimination ceased.

Every worker, however, had ten jobs to choose from that morning, among which the digging of holes was least seductive, and so the act did not go on. Not until Tuesday morning were the "appurtenances thereto" installed.

Since the act was an expensive one and we lost it on our biggest day, John would have appreciated some expression of regret on the part of the agent. "Appurtenances thereto" did not, indeed, provide a very detailed description of the Fair's responsibilities in the matter.

To this day, the mention of a high wire will elicit John's general advice and philosophy on the signing of legal documents.

"When you see a paper like that with all those sections in fine print that would take an hour to read and a lawyer to decipher, just make up your mind you're going to be wrong before you start. Sign it if you have to, but do it with your eyes open. No part of its contents will work to your advantage."

Thursday was a good, uneventful day, during which John was able to get some additional pipes laid to augment the water supply at Machinery Hall where the new artesian well was proving inadequate.

On Friday, all pandemonium let loose.

Friday is traditionally Danbury Day, when the native-born expects to be reunited with relatives and old friends whose lots have fallen in less-favored places. It had also been for many years the day of Fair Week when the schools closed and all children were admitted free of charge. This year the school board had voted to discontinue the holiday, since they considered with some justification that Saturday was equally good for fair-going. Had Monday's downpour been postponed until Friday, their nefarious scheme would no doubt have succeeded and set a new precedent. Friday, however, dawned clear and sunny and, oh, the difference to all concerned!

On that morning, the students went to school as usual, but many more than commonly did so carried parcels of lunch, and few showed any disposition to go inside. When, instead of heeding the bells for class, a majority began to line up in columns of four, it became evident that a movement to the fairgrounds was very literally on foot.

Starting from the high school on White Street, the marchers proceeded up Main to West, pausing at certain points to absorb reinforcements, since the talented organizers of the conspiracy had foresightedly dispatched pickets to the various grade schools with placards reading "Danbury Day," "Fire the School Board," and "Unfair to the Fair." Their ranks, swollen by many recruits of smaller stature, made their exodus from the city en masse, as if led by an invisible Pied Piper.

Warning of the impending invasion was conveyed to John by a flurry of telephone calls from points along the route of the insurgents. Soon after they started up West Street, he was able to reach Mr. Sweet. They had just passed his office, the superintendent stated, singing at the top of their lungs.

There was really nothing anybody could do by that time he and John agreed, since we were neither prepared to withstand an onslaught of two or three thousand, nor to use force against children. One gate was therefore opened for their free admittance and in they trooped as of yore.

All day they cavorted happily, rushing up and down the midways, racing through the Big Top, swarming over the rides, and no one said them nay. It is only just to their parents to presume that many homecomings were not so enjoyable. When the marchers are old men and women, they are all likely to remember that day for one reason or another, and to agree that not all lessons are learned from books.

The next year, the school board reversed their ruling, and now 7,500 Danbury children receive Friday passes from their homeroom teacher, which puts things on a more orderly footing.

Farmer John watches over thousands of people in a weekend crowd, 1969.

The remainder of the week passed more smoothly, but not uneventfully.

When somebody broke into our police station on Saturday night while the officers were employed on their rounds and stole some blankets a faker had deposited there for safekeeping, we were neither staggered by the atrocity, nor did we question for whom the bell tolled. We knew it tolled for us, as usual, even though no contract was involved.

The weather continued fine and our attendance for the last Saturday and Sunday passed 48,000.

On those warm October days, the park benches were in such popular demand that John kept writing memos to remind himself to add to their number in the future. The Fair, he discovered, is more than a showplace for cows, horses, sheep, goats and turkeys. It is a museum of humanity as well, where the contemplation of one's fellows constitutes an unlisted, but notable, attraction. Thus, a restful seat for observation purposes has a value of its own, not much inferior from anybody's point of view to the price of a grandstand seat.

For another year, John determined to create more free parking by removing obstructions and redeeming more swampland.

There wasn't much we could do to expand Connecticut's highway system, but the congestion could be partially relieved by opening future Fairs on Saturday instead of Sunday, thereby splitting up the opening attendance. The extra day would also furnish a sort of rain-insurance against a possible washout on the following vital weekend. Carrying out this plan resulted in our present nine-day Fair and had the hoped-for effect of making for happier relations with the public and with the state police. Instead of 42,000 patrons on Sunday in 1946, we had 11,000 on Saturday and 34,000 on Sunday in 1947.

16

CROWS IN THE CORN

Although experience is usually easier to get than money, its accumulation is not necessarily swift or joyful.

We were lucky to have had eight programs of midget races in 1945 and a whole summer of Saturday night racing in 1946 before our first Fair. In the course of these events, there were many hitches due to accidents, rain-outs and disputes among the contestants. There were parking problems and seating problems, each with its lesson for the novice. From these and other dilemmas, John emerged exuberant and with a better understanding of his job.

Less heartening, however, were certain episodes that revealed the hitherto unsuspected depths of chicanery among his fellows and new attitudes toward himself as Fair manager. It took a while for him to realize that he had somewhat abruptly become a public character and, as such, a natural target for promotion and criticism.

The first of these incidents were merely amusing. John came in after work one day humming a little tune. As he dumped the sand out of his shoes into the kitchen sink and prepared to enjoy his home, some words of the song became recognizable.

"Brightly the cuckoo is calling now, I hear him down under the hum-de-day."

I was unfamiliar with this composition.

"What is a hum-de-day?"

"Oh, I don't know any more of that. It's an old piece my mother used to sing. Keeps running through my head. It's so appropriate."

"Appropriate?"

"Sure. The woods are full of cuckoos this spring. I have a dozen crazy propositions a day. A fellow came in this morning and wanted me to go in on a deal to manufacture automobiles for the army. I just sat there and listened to him. Made me think of Amos and Andy."

"What kind of automobiles?"

"Some kind of improved jeep, I believe. We didn't get into it very far. I never saw the man before. He said he was just back from Washington and can get an order for 10,000 cars."

"How would you figure in the partnership?"

"I'm supposed to build a plant on the fairgrounds and furnish the money. He furnishes the idea. We could start with $100,000."

"I thought it took millions to go into automobile manufacturing."

"The first $100,000 would get us tooled up to build a couple of samples. Then it might take another $100,000 to get the design approved by the government. He would guarantee to get it approved."

"Must have influential friends."

"Yes, indeed, he was talking about it to Walter Chrysler in New York this morning. He just went into Chrysler's office, put his feet up on the desk and described the proposition. Walter said, "It's 100% sure-fire. It will make a millionaire of you, George. I'll help you get it approved."

"Why doesn't Walter put up the money?"

"Oh, Walter's too busy and has too much money already, but Walter said, 'George,' he said, 'You're a wonder. You have an idea a minute. All you need is capital.'"

"Walter hit the nail on the head all right. And what did you say?"

"I just told him, 'George, it's an opportunity all right, but I can't see working for the government. I've done work for the government, and they change their specifications too often. I'm afraid I'm not your man, George.'"

"Why you have to listen to everybody who steps into your office is beyond me!"

"I can't help it. They look all right and, as they talk, I keep thinking there must be more to this than I understand. By the time they get through, I'm wondering why they pick me out. Do I look foolish? It's a little embarrassing that way."

"Oh, they just canvas every prospect who will spare the time for them," I reassured him. "If you weren't so curious you could save yourself a lot of bother."

"Think so?" he said and off he went upstairs, but I had not heard the last of "Brightly the Cuckoo"—not by a long shot!

Not long after that, John advised me of an impending caller, a publicity man.

"He wanted to see me at the office this morning," John explained, "but I took your advice about not wasting my time. I was pretty busy and so I told him he could drop around tonight."

"Very well, I shall retire to the sun porch," I stated coolly.

At the appointed hour a knock came at the door, and I went into my voluntary exile, from which vantage point I saw a dark dapper salesman-type make a hand-wringing introduction of himself. John fears and hates hand-wringers. Why he refuses to wring right back I shall never understand, since he has large capable hands and could give better than he gets. With unimpaired cordiality, however, John took his caller's stylish coat and homburg and settled back to listen. He loves company of any kind.

"Mr. Leahy," the man began, "do you realize that thousands of people in this area who have never seen you know you by name and reputation? Do you realize what these people think of you, what picture they have of you?"

John smiled broadly.

"I've lived here all my life. I have a general idea," he said.

"I could build you up by modern publicity methods to be one of Connecticut's great men. My company would play you as a public benefactor. It's a great angle. Do you realize the secret of John D. Rockefeller's popularity in his declining years? He employed a press agent who hit upon a simple device calling for his client to hand out dimes to children. From being one of the country's most disliked men, Rockefeller became beloved by all."

"Money was no object to Rockefeller, I suppose, but it still must have cost him plenty to become beloved by all."

"Mr. Leahy, you know you are a very wealthy man."

"If I am, I'd better stop working and see the world. I've always wanted to."

"Now, a few thousand dollars wisely spent for a personality lifting such as my company could provide you would pay a rich dividend. It would be one of the best investments you ever made. I have been going over your general status and activities since I called you this morning. It is my considered opinion that you are worth close to a million."

A pause here for confirmation, John's affability seemed to be fading.

"That's what you were doing all this afternoon?" he questioned incredulously.

"Yes, and I feel that I have made a conservative estimate."

Another fruitless pause.

"Now my company's plan—"

Here, John interrupted.

"That estimate seems high to me, but, of course, if you make one every afternoon, it's not surprising. Anyway, I don't believe I'm old enough yet to want to be a philanthropist. Rockefeller waited until his declining years, you say. Leave me your card and I'll let you know when I reach that stage. I feel pretty well right now. Excuse me," he added, rising and risking his right hand again in a parting gesture, "My wife is waiting for me I'm afraid I'm wasting her time."

I emerged merrily from my retreat as the door closed on Mr. Slick.

"Oh, go ahead and get your personality lifted. Have it done while you are able to enjoy it."

"I must look foolish," John sputtered.

"Not necessarily," I encouraged him.

The encounters began to be less amusing, though, as they accumulated.

One Saturday noon, John was just finishing lunch when we had a caller. This time there was no appointment.

"I want to see John," stated a brash looking individual who stepped in as I opened the front door. "Hubcap is the name."

He was nobody I recognized, but the name was vaguely familiar.

"Sit down, won't you, Mr. Hubcap, I'm sure John can see you for a few minutes."

That ought to discourage any long sales talk, I thought. We were planning to get started for a dog show in Westport as early as possible.

"Hello, Harry," John called from the dining room, "Be right with you."

Harry tossed his hat on the sofa and began to light up a cigar. When I saw that no effort to put him at ease would be required from me, I decided to profit through this diversion by going upstairs to get dressed.

Half an hour went by before I went to the top of the stairs to take a sounding.

"It don't make a damn bit of difference," I heard Harry say, "I get twenty passes for the season or you don't get the permits. See?"

"You don't get twenty passes. You don't get any passes whether we ever get a permit or not, and I'll tell you something else—"

It occurred to me that this interview had gone far enough and I decided to run some interference.

"Hello, John," I called down, "I'm afraid we ought to be starting soon."

"I'm in an honest business," John continued heatedly, "and I'll still be in it the year you don't get re-elected. Furthermore—"

I descended upon them.

"Sorry to interrupt, but—"

"You can go straight to—"

"We are going to be late, I'm afraid."

"City Hall, and I'll be down there by the time you are and we'll see—"

"You can see how it is, can't you, Mr. Hubcap? John would like to spend more time with you, but we really have to hurry."

I cut across the line of skirmish to where the cat was dozing in a chair on the opposite side of the room. Scooping him up with a whispered word of apology, I cut right back again and dumped him out the front door, which I failed to close.

"What are you doing?" John bristled. "Can't you see we're not through?"

"That's all right. I just put the cat out. You'd only wake him up anyway with all that loud talk. He sleeps days," I added conversationally to Harry, "but believe it or not, at night he's really quite a ratter. You'd have to watch out for him if you came here after dark."

Harry seemed to have little aptitude for small talk, but he could see the jig was up and left without recourse to handshaking. John appeared more thwarted than appreciative.

"That so-and-so would take the buttons right off your coat," he grumbled, "but you have an awful nerve bursting in with that line of chatter."

"I'm sure Harry's been called a rat often and much more explicitly. Come on now, it's 1:30 already."

P. T. Barnum had known how to handle such emergencies. He kept a vest pocket full of printed cards that, at first glance, resembled tickets. These he handed out freely in response to requests for passes, but where a pass would ordinarily read "Admit One," these pasteboards were inscribed with a verse of scripture: "So he paid the fare thereof and went." (Jonah 1:13)

In Danbury, the tradition of "getting in free" was a carryover from the good old days when, not hundreds, but thousands of Fair passes were issued by the officials to heads of departments, who distributed them each according to his own discretion to employees, exhibitors and concessionaires.

Many townspeople "helped out" Fair Week in some capacity or other. They could ask for, and usually receive, passes for members of their families.

Then there were the stockholder's tickets, which were good for admission every day of the week. Rather than "waste" his pass on a day when he would be otherwise engaged, the holder might bestow it on a less fortunate neighbor. This gesture made for good fellowship, but it is easy to set a precedent and hard to recall a benefit conferred with the result that the following year the owner of the pass felt that he must offer it again to "good Old Elmer," whose hedge clippers he had been putting to good use. In turn, "good Old Elmer" was counting on the favor's being repeated, and making his plans accordingly.

It was indeed an improvident Danburian who approached the gate without some kind of free pass. To dig down for the price of a ticket was to reveal a lack of proper connections that was almost embarrassing. Rather than undergo such humiliation it was better to scout the ticket takers for an acquaintance who would be lenient enough to honor a matchbook in lieu of a ticket.

These facts were not unknown to the directors, who were the best fellows of all, since most of them ran competitive businesses or dabbled in politics. As to firing a ticket taker, that was out of the question. If he were not some official's relative, customer or constituent, he was sure to be a connection thereof.

Such a loose gate probably fostered the sense of proprietorship that our citizens have in the Danbury Fair, but it signally failed to provide any surplus funds with which to maintain the property over a four-year period of inactivity, such as the war caused. John was resolved to plug these leaks by keeping passes to a minimum and was duly braced to withstand some overtures, but he never expected the avalanche of requests, demands and importunities that poured down upon him as Fair-time drew near.

Early in September they started, but our first experience was rather encouraging than otherwise, since it demonstrated the esteem in which Danbury Fair passes were held.

The tickets had just arrived from the printers and I was busy in John's office extracting two rolls from a carton for presale, when in came my first customer. Elderly, but spry, he was with a smile and a handshake for both of us.

"Want to get sixteen passes to send out of town," he told John. "I was riding around over there today—see you've got things all shined up."

"Do you like it? Won't you sit down, Mr. Benedict?" John can always spare time to listen to that kind of talk.

"Nope, can't stop now. Goin' to the city. Want to buy sixteen passes. Don't want 'em for nothin'. I know you spent money over there."

Out came his wallet and he extended a $20 bill.

"My wife will get you the tickets. She was just opening them up when you came in. How did you like our new roads?" persisted John, to whom one taste of praise is just an aggravation.

I was putting sixteen adult tickets in an envelope when Mr. Benedict stopped me.

"Not tickets, I didn't say tickets, I said passes."

"We can't sell passes," I answered. "The law says the price and amount of the tax must be printed on each purchase."

"I don't care what the law says. I said passes and I meant passes. I can't send my friends tickets with the price printed on them. I've always had passes and I'm willing to pay for them."

"This is quite a new law, Mr. Benedict. Why don't you take the tickets?" I coaxed. "Your friends will appreciate them just as much."

"It's no honor to buy a ticket. Anybody can buy a ticket. Used to be I always sent passes to my friends. Didn't have to ask for them either."

He was showing such signs of impatience with my regard for technicalities that I feared for my first sale.

"I'll tell you what. I'll sell you the tickets and tear them now in your presence. Then I'll find you some exhibitors passes to send your friends. Now do you think Mrs. Benedict would have a piece of fancy work she would like to exhibit just for the fun of it?"

"She sure has. More 'n one way to skin a cat," he chuckled and went off toward the railroad station in high spirits.

Mr. Benedict's attitude was exceptional, however. In most cases, the honor of holding passes was secondary to the chagrin over parting with $1.20.

The phone would ring.

"Hello, Tom Torrent speaking. John, old pal, I'm going to need a dozen passes for Fair Week," a well-heeled executive from one of the hat shop offices would state.

"Passes," John would repeat dully, "what kind of passes?"

"Passes for the Fair, of course. They always mailed them to me early. I get them every year. What's the matter with your records?"

"Guess all the records got burnt up in the last fire. We're starting from scratch, you know. Anyway, there won't be any passes this year, Tom. The Fair needs a dollar from everyone who comes in order to pay some of its debts and keep running."

"Not going to give out passes! That's a hot one. You'll soon find out where that policy will get you. The Fair is a Danbury institution and Danbury people expect some consideration."

"Listen, Tom, you like the Fair, don't you? That's good will and it's an asset, but the Fair can't run on good will. Does the management in your factory give out free hats to Danbury people? Could they run the hatting business that way?"

"That's altogether different," was the indignant answer.

"Yes," agreed John, "it is. You have a whole year in which to show a profit. We have only a week. Try to look at it that way. We'd like to do it, you realize, but we can't."

Strange to say, the people who could best afford to buy tickets were the ones who regarded a refusal as a personal insult. John was kept busy explaining that as far as he knew only exhibitors, employees and the working press were entitled to receive passes. When an applicant continued insistent, John would reach for his pad and pencil. "I'm afraid your application is too late for this season's Fair, but if you will leave your name and address we will take the matter up at the next meeting of the board of directors."

By the telephone, a spindled stack of names and addresses was steadily increasing by five to ten calls an evening. It was almost unbelievable.

For years, the Fair's printing had been done in a small shop in Worcester by a man who had made the specially engraved plates adorned with a lot of old-fashioned scroll work.

John had taken some pains to find the printer's address and had written for an estimate on some work. A reply had been received promptly, but it was from the printer's widow, who said her husband had been dead for four years. His shop had been sold and the business discontinued. John wrote her again to see if the plates might still be in existence, but she answered that during the war she had given them all to a scrap metal drive and supposed they had been melted up.

It occurred to John that here was bad news that might be turned to good account.

"From now on I'm going to tell everybody that the man who used to print the passes is dead, and that's no lie. The light touch, you know. How does that sound to you?"

"Like a bad joke, and I don't advise it."

The bad joke, however, saved a lot of time and seemed to work as well as any long rigmarole of defensive explanation. It was really better not to dignify all these requests by serious treatment.

More disturbing were the trick-or-treat tactics of certain pressure groups whose demands for whole blocks of passes in the name of charity were followed too closely by threats of public vilification.

John is, I'm sure, naturally more charitably inclined than the average citizen, but being bulldozed into donating on an or-else basis aroused in him other than charitable instincts.

Why, he wondered, were these people so anxious to rush into print with their stories?

After a close scrutiny of the personalities involved, it dawned upon him that many an organization was headed by an opportunistic candidate for political office out to make a name for himself on some issue or other. If such an official could wangle the passes, he would be a public benefactor. If he failed, he would get credit for trying. He had nothing to lose, but his time.

Since John has little taste for the role of stepping stone, he has, at the risk of being considered a skin-flint, refused to give passes to any organization. To deserving groups, he prefers to make an outright gift of cash with no strings attached. To this policy he sticks, come hell or high water or adverse publicity. As for those schemers who hope to reach heaven riding the coattails of the under-privileged, if they should all be caught in down-draft, he feels it would serve 'em right.

17

GREEN PASTURES

The convenience of our home to the office was such that after fifteen years of marriage I had abandoned all hope of change and resigned myself to perpetual insomnia, when suddenly, in the spring of 1950, John came up with an astonishing proposition.

"How," he inquired diffidently, as if knowing me mortally opposed to change, "would you like to move to the fairgrounds for the summer?"

How would I like to move to the fairgrounds? How would I like to get away from the noise and dirt? How would I like to plant a garden? How would I like to go to heaven when I die?

"How soon could we go?" I asked him, eager to clinch the matter then and there. It was May already and no time for toying with ideas. Until that moment I had fully expected to live out my days on White Street. Now I could see myself transplanted to the green pastures and still waters of the fairgrounds for six months out of every year. It was too good to be true. I felt like Proserpine.[56]

It was no flash-in-the-pan either. The self-same compulsion of John's that had tied us close to the office year after raucous year now summoned us to the scene of his latest interests and exertions.

My deliverance was the indirect result of a change in the public appetite for midget racing.

In 1949, during the fourth full summer of that sport, the crowds had begun to taper off. Ours was not the only track so affected. It just seemed that a lot of people everywhere had little-by-little discovered other ways to spend Saturday evenings.

Our regular attendance by September had boiled down to about 2,000 bona fide racing fans, at which point the weather became altogether too important. A sudden shower just before or during the first four events would either wash out the program completely or cause it to be rain-checked, converting a potential slight profit into the loss of our heavier fixed expenses. The enterprise was becoming a bit too speculative, John thought, even for him.

After that year's Fair was over and put to bed, he had time to go over the records of the past racing season and to talk with members of the American Racing Drivers Club who knew conditions at other tracks. He had about reached the painful conclusion that fashions in amusements are subject to change when there appeared to him, meditating and conferring, two gentlemen straight off Broadway with plans for the summer 1950.

They were Ben Boyar and James Westerfield, who had been casting about the exurbanite section of Connecticut for a site upon which to establish a summer theatre for the production of popular-priced musicals. Both were men of considerable theatrical experience. Mr. Boyar was general manager for Max Gordon at the time. Mr. Westerfield, by his own billing, was "the West Coast entrepreneur who founded the noted outdoor 5,000-seat theatre in Los Angeles."

The theatre in the round, or music circus, had been introduced with apparent success by St. John Terrell and Lawrence Schwab in Lambertville, New Jersey, and it was this form of production which Messrs. Boyar and Westerfield had in mind. They planned a series of twelve operettas, alternating those of the Romberg, Friml and Herbert variety with musical comedies of more recent vintage. A regular company of outstanding singers would be reinforced by a different visiting star at each presentation.

Since buying a site and paying for a tent with a capacity of 1,500 seats, as well as a second more substantial shelter for dressing rooms, wardrobe and toilets, would leave them with scanty capital for operating expenses, they proposed to

56 A Roman goddess based on the Greek Persephone and her mother Demeter, the goddess of grain and agriculture.

rent the fairground's Big Top on a weekly basis.

The Big Top was a natural for the project. It could accommodate at least 2,000 seats for a theatre in the round set up. In it and the adjacent Administration Building were all the necessary facilities, with the exception of a center stage, stage lighting and sound equipment.

Contralto Marian Anderson, the first African-American woman to sing at the Metropolitan Opera, also graced the Danbury Fair stage in the 1950s.

Now, John will listen to almost any kind of music, but he has a special weakness for the sentimental melodies of yesteryear. He was consequently favorably inclined toward the undertaking from the start. A rental price was soon agreed upon and the name "Melody Fair" chosen for the series.

Plans were going forward for a June opening and John was zestfully readying the grounds and buildings when, halfway in the course of his preparations, certain problems of maintenance began to bother him.

Who, for instance, would close the exits after the show was over? Who would turn off the lights, set the rubbish containers outside and otherwise protect the premises from the haunting hazard of fire? Our caretaker at the front gate had a splendid record of reliability, but only his own physical presence, John decided, would insure meticulous nightly inspection.

There was the farmhouse standing vacant, except for Fair Week, as it had since wartime when Henry Johnson moved away. Perhaps he could persuade me to break with convenience and come to live in it for the summer.

"How soon," I had answered, "can we move?" and we went over to see the house.

It had been serving as headquarters for the antique show for four years, and so I had been through it a few times, but never with an interest in its livability. Now I went over it with a seeing eye. It was a high-posted Victorian creation with big rooms and long narrow windows. The stairs were steep with high risers. I recognized it at once for what is known to Mainers as a "woman killer."

One woman, perhaps, yes. Two women, no. Mrs. Keating, who had stood by us for upward of ten years, was willing to participate in the adventure.

Actually, the house was a pleasant one. It was elm-shaded and had a long white-columned porch on the side facing the meadow, which lay between it and the horse barns. A huge upstairs bedroom that I quickly earmarked for myself had French doors opening on an upstairs balcony.

It was on the fifth of June, John's birthday, that we moved. About two weeks later, Melody Fair presented its first musical, "The Merry Widow," with Irra Petina[57] as guest artist. Advance publicity had been thorough and the season got off to a flying start with a well-filled house. Local newspaper comment was extremely complimentary and even the New York critics had good words to say. On the front page of its amusement section, the Herald Tribune ran a long friendly appraisal with sketches of the physical layout by George Shellhase.

The second operetta, "The Chocolate Soldier," was another hit, with the verdict favorable not only in the newspapers, but again in the box office where it counts the most.

After such an auspicious launching, the company settled down to regular hours of rehearsal and the extracurricular enjoyment of summer in the country. Actors stalked past the bay window of our dining room practicing their vocal scores with appropriate gestures while dancers performed their acrobatics on the lawn.

Romances sprang up overnight and flourished under the elms or on the shady side of the big water tank or two steps off the beaten path in any direction. Couples studied their scripts together or stretched out in the sun, pillowing their heads upon them, as if hopeful of absorbing their contents by osmosis. Two especially delightful members of the ensemble found time for a church wedding on their Monday off. Other love affairs, more temporary, but no less fervid while they lasted, pleasurably electrified the atmosphere. There were spats and makings-up and more spats.

The whole company was united and constant in its affection for Henry, an adolescent Rhode Island Red, which either ran away to join the show or was chick-napped as a mascot by some spry young member of the cast. Whatever his origin, he quickly adapted himself to show business and grew very choosy about his diet, always preferring to wait for the tail end of a hot dog rather than bother with the cracked corn that was lavishly provided. His voice was showing constant improvement both in quality and volume, until one night when he suddenly crowed during a performance, upsetting a skittish maiden on the top row of bleacher seats where they both were roosting. Her downfall was Henry's, for

---

57 An actress and singer, as well as a leading contralto with the Metropolitan Opera in New York City, called the "floperetta queen" by critic Ken Mandelbaum.

Mr. Boyar banished him to the nearest farm. There he must have found life flat, stale and poor pickings after his career in light opera.

Unfortunately, attendance at succeeding performances failed to fulfill the promise of the two early shows.

Irv Jarvis plays the piano in an impromptu singalong with cast members from a Danbury Fair operetta, 1950-51.

The third offering, "Anything Goes," was not well patronized on weekdays although Saturday and Sunday evenings continued to hold up. That pattern persisted thereafter except for a heartening boost in attendance in the seventh week as Bridgeport and Waterbury turned out in force to applaud Mimi Benzell[58] in "Naughty Marietta."

In spite of good quality, which was pretty well sustained throughout the series, and low prices, which ranged from $1.20 for the more distant of the bleacher seats to $2.40 for a ringside yacht chair, it was evident sometime in August that the success of the venture would not extend into the realm of finance. On September 10, the season ended on a slightly sour note for which no vocalist was responsible. For awhile there was talk of a "safari for angels with capital to finance a second attempt," but that idea was gradually abandoned. John had become more deeply concerned with the success of the venture than his position as landlord warranted. Busy as he was all summer with preparations for the Fair, he had drawn freely upon his time and experience in cooperation with the producers. He was loath, indeed, to

The cast of Cinderellaland, one of the Fair's fairy tale areas, 1965.

58 An American soprano who hailed from Bridgeport, CT, and performed with the Metropolitan Opera before establishing herself as a Broadway musical theatre, television, and nightclub performer.

believe that such popular entertainment could not be built up to pay its way. He had a good mind to try his hand at it in 1951.

The only feature of these musical evenings that he had not particularly enjoyed was poking around the Big Top with a flashlight every night after the show was over. There, his self-imposed duties of night watchman had revealed a horrifying nonchalance on the part of city people toward the lighted cigarette butt.

Top: Cast members of the Danbury Fair operettas assist in righting a toy soldier as John Leahy looks on appreciatively. Bottom: One of the operettas being performed in front of the grandstand, 1951. Note the Big Top canvas stretched from grandstand roof to the rear of stage.

"Why," he wondered, "couldn't the concrete grandstand be used as an auditorium?" With a stage house in front and canvas stretched between the proscenium and the edge of the grandstand roof, the actors and audience would be protected in case of rain.

In the nearby Silvermine district of Norwalk, there lived a man who had pioneered the summer musical in the '30s. His name was Greek Evans.

His "Theatre-in-the-Woods," inaugurated as part of the artistic tradition of that locale, had thrived for six years, drawing big audiences from the well-populated towns along the Post Road.

Whatever caused its final dissolution was not a lack of energy or experience on the part of Mr. Evans and his wife, a former Metropolitan artist, Mme. Henriette Wakefield. They had weathered many a gale. After it closed, Mr. Evans demonstrated that in addition to being a talented singer, director and producer, he had the makings of a first-rate carpenter. On the large tract of land that he owned, he began to build houses, doing so well as times improved in the real estate business that he had never attempted to revive the theatre.

Notwithstanding his successful invasion of the business world, Mr. Evans was by nature and training of the theatre, by the theatre and for the theatre. He impressed John, who went to him for advice, by his sincerity and enthusiasm as well as his practical knowledge of stagecraft.

"I don't suppose," John suggested, "that you would want to direct the shows?"

Upon reflection, Mr. Evans decided he could spare time for the short season that John planned.

Since this undertaking was to be in the nature of an experiment, it was thought best to attempt only six operettas, each presented three times only, on Friday, Saturday and Sunday nights.

A stage with ramps was set up in front of the grandstand and roofed with green-striped canvas. John obtained some excess pieces of scenery from the Metropolitan opera and supervised the installation of lighting and sound fixtures. He saw to painting of three sections of grandstand seats in red, yellow and blue to match the differently priced tickets, and he adorned the seats with foam rubber cushions in appropriate colors. He had the parking lots clipped till they resembled fairways and set up, as directional guides, half a dozen gaily painted 12-foot soldiers with right arms outstretched to point the way. He even provided an ironing board to the somewhat exacting specifications of the wardrobe mistress—who ever afterward brought him her numerous complaints.

With these and a thousand other details attended to, he sat down to enjoy the opening of "The Student Prince" on the evening of July 20.

Again the Herald Tribune spoke kindly of the enterprise, but as "New Moon," "Firefly," "The Merry Widow," "Rose Marie" and "Blossom Time" were presented during the weeks that followed I, for one, was in no position to offer any critical judgment. I had been assigned to the box office with two charming young ladies as assistants, but as usual with no experience to guide me.

Short as the season was, it was long enough to cure John of all theatrical ambitions. After the nuts were out of the shells, as he likes to remark, the only profit to anyone was vested in that super ironing board, which I inherited and continue gratefully to use. My long-term gain was more substantial. After two summers on the fairgrounds, John couldn't understand why it had taken me so many years to recognize the advantages of country living. Now we move in April and October.

John clings to the idea that, with time and plenty of money, the summer musical could be built up in this vicinity, but he feels it is now somebody else's turn to try. He will be a regular patron.

Vain regrets, however, have no place in our scheme of things. We plunge into and out of experiments like sea lions in a pool. Long before the pussy willows in the swamp had opened, we were ready to try another form of entertainment. It was auto racing again, but with a difference.

While we had been otherwise engaged, the tracks at Bridgeport, New Haven, Stafford Springs, Plainville, Avon and Rhinebeck had changed over from midgets to stock cars, with surprisingly good results. Stock car racing was a far easier and cheaper amusement to promote than midget racing with its $15,000 open Offenhausers. It was also far less dangerous. The stocks in use locally were not new cars, but old coupes and sedans, somewhat souped up beneath the hood and plentifully braced and padded inside the body, so that by adding pontoons a daredevil might survive a trip over Niagara. The reinforced steel sides and tops protected a driver in the frequent side-swipes and roll-overs that were a part of the proper enjoyment of this sport.

While the driver's skill was important in winning a race, it was not such a life-and-death matter as with the midgets. A driver could take a few chances, with the result that less experienced chauffeurs occasionally won races when their cars held together and stayed on the track. In nearby Brewster, New York, a group of youths calling themselves the "Southern New York Racing Association" (SNYRA) had improved a track out of a somewhat level field. They had not been able to provide seating or other customary conveniences, but that did not deter a substantial number of standees from gathering, dressed in old clothes after the initial experience, to watch the dirt fly on Sunday afternoons.

John had visited more conventional plants, where the hard surface one-fifth mile tracks were being used for stock car racing, and had not been impressed by the sport. Besides being too small, such tracks lacked a certain springiness that improved the quality of racing at Brewster. Then, too, there was something spectacular in the sight of flying dirt until it became transformed into dust clouds. SNYRA was putting on the better shows, he thought.

In February, the officers of this organization met with him and Mr. Jarvis to arrange a union between our facilities and their club. Outside our midget track lay the abandoned waterway, dry and empty, its bed lying about four feet below grandstand level. Filled in and widened, it would make a one-third mile course. With a light application of clay to the straightaways and a heavy one on the curves where skidding tires are apt to gouge holes, the dust problem could be overcome. In the interests of the public, it would be best to leave the homestretch unfilled so that its depth might provide additional protection from the sociability of the stocks, which have been known to somersault fences into the laps of spectators.

The amount of dirt-moving necessary to build the new track led to the purchase of a road scraper and a new tractor. We already had a bulldozer and a pay loader, which were well on the way to paying for themselves.

In June, the races got off to a warm reception, in which the presence of locally known drivers was more of a factor than anybody's guesswork had forecast. Before long, the backyards and maple-shaded streets of this locality were spotted with vividly painted vehicles bearing huge black numerals. SNYRA was besieged with applications for membership and we were quite plainly embarked upon an era of spills and crashes.

There were devotees of midget racing who looked with disfavor upon the new diversion, condemning it as a thrill show in which skill and experience play minor parts. That those characteristics can be acquired became evident as time went by. For a change, and for the time being, the public liked it better.

A certain phrenologist[59] with whom I am on terms of easy familiarity, and vice versa since he puts in his appearance annually at Fair-time, once told me that stock car drivers have a depression where the bump of caution ought to be.

"Let me show you," he volunteered, and I bowed my head for examination.

"Well, it ought to be right there," he faltered.

"Right where? I don't feel any bump," I commented skeptically.

"That's funny, neither do I," he admitted, eyeing me with awakened interest that I assumed was professional. At least, there's no depression.

Whatever other charges may be brought against stock car racing, it remains, as John succinctly puts it, "a hell of a night's entertainment for $1.20."

Racing Through the Years at the Danbury Fair

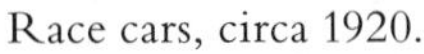

Race cars, circa 1920.

Auto racing action.

Race car lineup. *All photos this page, Frank Baisley*

59 A person who studies the shape and size of a skull as a supposed indication of character and mental ability.

Ira Vail in his #4 Duesenberg, 1914.
He raced at the Indianapolis 500 several times,
but never won. *Frank Baisley*

An early motorcycle race. *Frank Baisley*

Stock car pile-up, circa 1960.

Modified stock cars ran faster and crashed harder, 1977. *A. Cor-Kaptein, Jr.*

18

BRASS TACKS

"Say it's not easy," John admonishes me from time to time, fearful that my girlish enthusiasm will paint too rosy a picture.

He needn't worry. I know most of the facts of life as they apply to this merry-go-round of ours.

The Fair is not, as might appear to the casual observer, a nine-day project that makes quick profits and silently steals away until another season. From the Sunday night when it closes until the next opening date, its operation demands constant attention, supplemented by lots of good judgment and considerable physical stamina.

Putting it to bed starts early Monday morning when, weather permitting, the Big Top is taken down before some skittish hurricane decides to veer in our direction. Wind is a slasher of fabrics, as fairs and tent shows in New England have come to realize during the past decade.

Our canvas, which covers a space 246 feet long by 120 feet wide must be put away thoroughly dry or it will mildew. The weight of its sections requires the use of the payloader for their transfer to the concrete cattle barn where they are stored for protection against fire. The present flame-proofed Big Top cost $12,000 in 1951 and its life expectancy is six or seven fairs.

Top: Giant candy canes.
Middle: One of many fairy tale attractions at the Fair, the Old Woman Who Lived in a Shoe. *Danbury Museum and Historical Society*
Bottom: Daniel Nason steam locomotive and tender in New England Village.

For a week or two, John works with a crew of about thirty men picking up the pieces, putting away exhibits and storing the bulky fairgrounds ornaments. The whole transportation show, including the Daniel Nason, one of the earliest wood-burning locomotives, the old railway coach, the horse car and the Conestoga wagon must also be towed away to shelter. The giant candy canes, the 12-foot toy soldiers and all the decorative figures from Mother Goose and other children's stories must be carefully handled and housed. Then the grounds are raked clean and a mountain of rubbish trucked to the dump where it is covered with earth and bulldozed over.

To do such jobs, the Fair owns two tractors, a bulldozer, a payloader, a heavy truck and two pickups that, together with a road-scraping machine and two water wagons for work on the dirt track, represent an investment of over $26,000.

By the fourth week in October, we move back to White Street, but while the weather holds good John continues to make such changes and improvements as he thinks will save him time in the spring.

When he finally decides to give up outdoor work and become an office executive, he spends hours with Mr. Jarvis comparing notes on the success or shortcomings of each department with the purpose of improving next year's operation. Their conferences with each other and with the various department heads carry on well into the months that follow.

The Saturday after Thanksgiving finds John bound for the convention of the National Association of Fairs in Chicago, where he is likely to share his experiences and convictions in a speech or two. He also listens and learns and has fun enough to last him well into the joyous season just ahead.

Even as the snow lies round about in January and February, exhibits are being arranged for, attractions are being booked and concessionaires are being assigned space.

April means that carpenters can resume their pounding and painters their scraping and brushing. Then for a few weeks John takes his lunches in a brown paper bag, but as the world warms up in May and the races are ready to start, we move again, which pleases both of us.

June and July are busy, but beautiful, months. A cool breeze funnels through our level valley even in haying weather and there is no sense of being pressed for time. Once a week. our social life takes a spurt as we get dressed up to entertain our paying guests, the racing fans. Then, of course, bedlam is let loose for a few hours, but on other than Saturday nights, we can tell when the cat comes in by his noisy tramping across the kitchen floor.

As we gather speed in August, the pace of preparation grows hectic and remains so until Fair-time brings its customary climax.

John calls it a nine-day business with a year-round overhead. The avowed aim of the country fair is the advancement of agriculture. What state legislatures have in mind when they appropriate and allot money for premiums is improvement of quality. Quality has its innings at the judging of livestock, vegetables and fruit, at the awards for farm machinery, and at the horse and ox drawings.

A fair, however, cannot live by husbandry alone. None ever did and, today when so many people are more familiar with factories and offices than with farming, there must be, coupled with nobler purpose, the promise of a general good time. Thus, while the spirit of the Fair is competition, its atmosphere is that of fun.

The average fairgoer, while he enjoys staring at the prize bulls, the big pumpkins and the little pigs, is little conscious of the hard work and lofty aims of those who produced these phenomena. His interest in agriculture is as casual as mine is, for instance, in football, which under the most favorable conditions I sometimes find agreeable to witness. He comes to the Fair as he goes to the circus, to be amazed and amused.

Our patrons, made critical by such easy access to the metropolitan area, may be a little harder to amuse than a similar crowd in the Midwest or the South. At any rate, the newer and more varied the attractions, the better they are pleased.

The "Parade of Champions" cattle review took place each Governor's Day. *Frank Baisley*

We feature, therefore, the kind of grandstand show that is most popular at the moment along with spectacles designed to please the greatest number of people. Stock car racing is an example of the former and the auto thrill show of the latter. We do not attempt to present revues or musical shows, as some fairs do, because such entertainment goes best under lights, but loses its glamour in the bright sunshine of afternoon. Certain attractions, such as the Wednesday parade of cattle, hold perennial fascination, but many, like bicycle racing, have had their day and ceased to be in accordance with public taste and demand.

Even bicycle races may not be gone forever. The Wild West Show that went out with Buffalo Bill is again back in favor, thanks to Roy Rogers, Gene Autry and a lot of western movies. Buck Steele's "Frontier Days" was thus a very popular feature of our last three Fairs.

In highest esteem, as exhibits, are the very old and the very new, the very big and the very odd.

In these classifications come the first telephone switchboard, the General Electric show, the biggest hog in the world and Henry, the educated Brahma bull.

At present, one of our biggest attractions is the daily parade. Recalling the old-time popularity of the circus street parade, John conceived the idea of a parade around the midways and in front of the grandstand of horses, elephants, band wagons and floats, including almost everything on the grounds that can walk or roll on wheels. It has grown longer and more elaborate each year and, judging by the way its route is always jammed with spectators, it seems to have universal appeal.

Another great attraction is that Grand Champion, the Governor of Connecticut. We set aside Wednesday for him each Fair Week, and he sets it aside for us, regardless of politics, weather or prior claims upon his time. We are hard on governors in this state. After Gov. Baldwin in 1946, Gov. McConaughty[60] in 1947, Gov.

60 James L. McConaughy (1887-1948) served as Connecticut's governor from 1947-1948.

Shannon[61] in 1948, Gov. Bowles[62] in 1949 and 1950, it was positively restful to settle down to Gov. Lodge[63] for four years. Governor Ribicoff,[64] elected in 1955, seems destined to be with us for an equal period, and again this fall the male citizenry will foregather to shake his hand, while their ladies observe the governor's lady, how she looks and what she wears.

Governor and Mrs. John Dempsey review gifts from farm exhibitors during one of their visits. *Danbury Museum and Historical Society*

In an election year, we are deluged with candidates for office and John, feeling that the public is entitled to see them, usually introduces them on the grandstand. At the same time, by way of keeping the Fair out of politics, he tries tactfully to discourage speechmaking.

In 1950, Senator Bill Benton[65] was with us almost daily, dropping in by helicopter. He rode in the hayride around the grounds, tossing Benton buttons to all and sundry with such abandon that some are still being retrieved from the shrubbery by the fussier of our rakers-up.

After his appearance on the opening Saturday and Sunday, John, surprised to see him again on Monday, exclaimed after greeting him cordially, "But this is an off-day, Senator."

"That's all right," laughed Benton, "I'm only following instructions. President Truman told me to speak to, shake hands with, and look into the faces of 25,000 citizens of the state. He said that's the way to get elected and this seems like a good place to meet people."

The theory was apparently sound, for Benton's strenuous campaign plus this expert coaching resulted in his election. While the rest of his party, including Gov. Bowles, were swept from office in the Republican landslide of that year.

---

61 James C. Shannon (1896-1980) served as Connecticut's governor from 1948-1949.

62 Chester Bowles (1901-1986) served as Connecticut's governor from 1949-51.

63 John Davis Lodge (1903-1985) served as Connecticut's governor from 1951-1955.

64 Abraham A. Ribicoff (1910-1998) served as Connecticut's governor from 1955-1961.

65 William Burnett Benton (1900-1973) served as a U.S. Senator from Connecticut from 1949-1953.

In 1952, Senator McCarthy, no particular friend of Benton, apparently under the impression that Connecticut needed setting right on certain subjects, came to speak in Bridgeport and incidentally, let us assume, paid the Fair a visit.

It takes more than tact to discourage this campaigner, who managed to speak, not at length, but with quite a few partisan words, which brought us over the next few days many irate phone calls and a spate of letters denouncing our innocent grandstand as a political forum.

Again, as John says, "It's not easy," and completely unforeseen events may give use to a big batch of the kind of publicity we'd be better without.

Good drawing power is our greatest asset, since a self-supporting fair has only two sources of income, gate admissions and receipts from rentals.

For attendance, the Fair has to depend upon advertising and its reputation as a well-conducted enterprise. Proved ability to give each age group a good time in clean, pleasant surroundings is probably more important than advertising, although the public should be well informed of the dates.

Gate receipts must be protected by the realization that tickets are money and passes are money. It is better to hand out dollar bills, John says, to seekers of free admission, than lightly to give away passes. In the first place, they cost less and, secondly, the recipient gets a better idea of the value of an admission ticket.

For rentals of space and privileges, the Fair again depends on past performance. If a manufacturer knows that his product will be seen in operation by upward of 150,000 prospective customers, he wants to bring it here.

Although all our available space is usually contracted for well in advance, our policy is to keep rentals low as compared to most fairs. This leads to less dissatisfaction when bad weather reduces expected profits and to generally better cooperation with the management.

Within the circle of the Big Top lie the choicest rentals for the sale of goods. Almost every conceivable kind of merchandise is offered here, and some that is inconceivable. There are bandana kerchiefs and cowboy holsters, dahlia bulbs and African violets, honeycomb and maple sugar, lazy man's ties and Danbury hats, trick cabbage shredders and potato peelers, and toy monkeys running up a stick.

Outside are the food concessions, the cider and soft drink pitches, the weighing machine and the various stands of souvenirs, to name a very few of the many spots that contribute both income and atmosphere.

If a concessionaire is disappointed over his volume of sales, Mr. Jarvis analyses his problem to avoid another such debacle. He tries to select the spot best suited for each article and to avoid renting too many spaces for the sale of one commodity, with the result that in this department there is remarkably little honest cause for complaint.

Except during Fair Week, John doesn't offer any part of the grounds for rent. While rental agreements from before his ownership are honored, no new arrangements have been made even though requests from local clubs and organizations are occasionally received for use of the premises for picnics or other celebrations. Some offer to pay nominal rental, while others feel the facilities should be readily available for public use.

While reluctant to refuse, John has always been doubtful, chiefly because of the fire hazard, of the wisdom of having assorted groups of merrymakers roaming the grounds and buildings.

Once early in his management and against his better judgment, he did agree when a representative group of middle-aged citizens persuaded him to let their organization hold a clambake there. On the afternoon of the occasion, he visited the scene of revelry, and again took a look around after the party was over, to see that all fires were out and that no stragglers were locked in when the gates were closed. Despite these precautions, when he went over the following morning with the cleanup detail, he discovered a couple of the brothers stretched out unconscious in the hay of the oxen barn.

One of them would have been sufficient to convince him that it was foolish to sign large checks for fire and liability insurance with one hand, while extending hospitality with the other to fraternities which, however noble in conception, included such irresponsibles among their membership. Just thankful that their vice was not smoking, John unsympathetically added them to the litter of clam shells and empty beer cans on the dump truck for delivery into town.

It is not always easy to keep the barns filled with prize cattle for nine days, which is a long run for a fair. Many farmers feel understandably that they cannot spare their help for that period.

John's position, too, is understandable. He cannot allow cattle tie-ups to stand empty on either weekend when the heaviest attendance is expected.

The difficulty has been solved to a certain extent by the payment of fixed amounts to each exhibitor to compensate him for the time lost.

Sometimes when an owner knows we are especially in need of his particular breed his price rises above the usual sum. Then John balks.

One evening after a lengthy toll call to one such individual, John telephoned Mr. Jarvis to report results. I listened with sympathy and amusement to the one-sided conversation.

"He was willing to bring five Angus for $500." Pause.

"No, he can't send his man. Our help would have to feed and groom them. We provide the feed anyway." Pause.

"No, he wants the best spot for them, and he wants a blue ribbon for sure."

Another longer pause during which the receiver made with gay staccato crackling.

"No, no," said John. "I didn't tell him where to go. I told him I'd let him know."

But that was the end of the negotiation.

Not even dirt is cheap any longer.

Every season, John reclaims some more land from the swamp for parking. For several weeks trucks are busy carting, dumping and leveling. He gets fill wherever he can and as cheaply as possible, but the price always seems exorbitant for dirt. Since automobile manufactures continue trying to outdo each other in the size of their product, there is probably no end to this round-the-rosy.

Trotter with sulky and driver, 1934. *Frank Baisley*

Now to face a few facts about horse racing, the revival of which, judging from comments and letters received from many older patrons, would do so much to improve the Fair.

It was no grudge against horses and trotting that caused John to close the grounds and stables at the time when he and O'Hara said their emotional goodbyes. He had hoped and fully expected trotting to take its old place as the prime attraction it had been for so many years.

After the war, however, the popularity of harness racing in New York State was rapidly swollen by the yeast of pari-mutuel betting.[66] The rise of Roosevelt Raceway, less than sixty miles away, not only afforded horse lovers ample opportunity for seeing the best trotters under the most favorable conditions, but transformed tens of thousands, who had formerly thought a whiffletree[67] grew on a windy hill, into ardent converts to the sporting life. In 1952, another great raceway, still nearer, opened at Yonkers.

Connecticut so far has failed to legalize betting, which was once passed by the legislature, but vetoed by Gov. Wilbur Cross. This state of affairs recalls a somewhat similar situation before World War I and the Volstead Act,[68] when local option prevailed in Massachusetts. One city would be bone-dry while an adjoining municipality doubled its normal sales of liquor. An election slogan proposed by some wiseacre newspaperman went: "Prohibition for Boston and Rapid Transit

---

66 A form of betting in which those holding winning tickets divide the total amount bet in proportion to their wagers, less a percentage for the management, taxes, etc.

67 The pivoted swinging bar to which the traces of a harness are fastened and by which a vehicle or implement is drawn.

68 The informal name given to the National Prohibition Act, which established prohibition in the U.S. as of October 28, 1919.

to Chelsea." With plenty of gasoline again in their tanks, Danburians have rapid transit to their heart's content. It takes big money to make the mare go these days and the Fair cannot compete on such terms.

The remarks in the *Official Report of the State Board of Agriculture* for the year 1887 on the subject of running a successful fair hold just as true today. That recipe, if anyone recalls, contained three principal ingredients: plenty of money, the right kind of management and good weather. Money and management will, at best, make unleavened bread unless that third nerve-wracking component, good weather, is included.

One rainy weekend, and the Fair might break even. Two of them, we are in the red for another whole year. Two such years and the result is too painful to contemplate.

In 1953, the New Haven Railroad resurrected the railroad excursion from New York City. The next year, tickets were printed to include way stations between New York and Pittsfield, Massachusetts, with admission to the Fair included in the price of the ticket.

Whether this will be their policy in 1957 is hard to say, but railroad travel remains a convenient way of attending, with no traffic or parking problems.

From points less directly served by the railroad, there are many bus excursions and connections.

Weekdays are easiest if you plan to drive. If you come via the Merritt Parkway, leave it at Norwalk for Route 7 to Danbury. You have only seventeen miles more to go. Of course, Route 7 is pitted with antique shops, but do not let that deter you. The fall foliage should be at its best.

Some things are not perfect yet, but we are working on them.

Hi, Ho! Come to the Fair! And come, if you can, in the morning!

PART 2:
AMERICA'S MOST UNIQUE FAIR

by

John H. Stetson

Containing among other things
A Personal Reflection from
John Leahy's Step-Grandson
As well as a Continuance of
The History of the Great Danbury Fair
From the Mid-1950s
Until Its Demise

"To every thing there is a season, and a time to every purpose under the heaven.
A time to weep and a time to laugh; a time to mourn, and a time to dance."

Ecclesiastes 3:1 and 3:4

20

GROWING UP WITH "UNCLE JOHN"

My memories of John W. Leahy and the Fair started not long after my birth in 1944. My father, also named John, came to Danbury as a young man from his home in rural Mechanic Falls, Maine, shortly after his mother, Gladys, wed John. He went to work for the Leahy establishment in 1937 and was granted several military deferments from the World War II draft as he was working in the machine shop helping to fulfill wartime government contracts. The Leahy machine shop was making .30 caliber bullets, among other things.

During this time, he met my mother, Maxine, who was staying with a friend a couple of doors down from the Leahy home on White Street. She was from Philadelphia and was visiting her brother, Sid, who was working at a farm on the property known as Terre Haute in nearby Bethel. Maxine and John soon married, then his deferment expired and he was drafted into the Army. He was posted as a military policeman at Fort Sam Houston in San Antonio, Texas. Shortly before his unit shipped overseas, Mom became pregnant. She traveled by train to stay with her parents in Philadelphia, where I was born in September 1944.

After the war, Dad and Mom settled into an ancient farmhouse at the corner of Kenosia and Backus Avenues once known as the Harvey Backus Farm. The junction was known as Backus Corners, and it eventually became part of the Danbury Fair property. This was the first home I remember although it's long since gone. A Toys"R"Us store covers the spot now.

The old house was heated by a coal furnace. I can still remember a green dump truck backing up to an open cellar window and hearing the roar of the coal as it slid down the metal chute into the cellar. My mother, who was used to the relative luxury of her home in Philadelphia, talked about rats the size of housecats that played on the front porch. We had no car, but Gladys would take my mother to the grocery store or for other errands. My father was employed as a propane gas serviceman and installer for the Jewel Gas Company. The only fringe benefit he received was that he was allowed to bring the service truck home. Leahy employees worked a 60-hour week in those days, but so did a lot of other folks. Such was life for John Leahy's stepson and his family.

Uncle John,[69] never having had kids, was delighted with having me as his "nephew." (He was really my step-grandfather, but that is an awkward sobriquet.[70]) As a family, we celebrated every holiday at the Leahys' house. As soon as I was old enough, perhaps at around six or seven, I would stay overnight with them in the summer in the big old farmhouse in the center of the fairgrounds. I was allowed to "go to work" with Uncle John during the day as he traveled around the grounds in his black 1935 Ford coupe checking on the various work crews that spent the summers maintaining the buildings and making improvements.

In the evenings after supper, John, Gladys and I would take strolls around the grounds while John took in the day's progress and mapped out future work. As lightning bugs flitted about and woodchucks raced for their holes, John was jotting down notes on the white 5x8 pad he always carried with him. Every action John took had a work-related purpose, even our pleasant evening walks.

I was only two years old when John ran his first Fair in 1946. A few years later found me at his side during every day of Fair Week that I wasn't in school. He made sure I rode in the big parade everyday seated between Irv and himself in an elegant restored horse-drawn carriage, waving to the fairgoers and concessionaires. On weekdays, the crowd consisted of a high percentage of local people,

69 For a variety of reasons, John and Gladys decided to pass off her son as her brother. As a result, my siblings and I came to know our grandmother and step-grandfather as "Aunt Gladys" and "Uncle John." That little subterfuge was never admitted to by either one of them, although I became aware of it quite by accident when I was about twelve years old.

70 A nickname given to a person.

and John and Irv had a great time acknowledging them with hollered greetings. After my sister, Susan, was born, John actually had the two of us ride in our own carriage with our names emblazoned on the seatback.

The Wednesday of each Fair Week was traditionally Governor's Day. Danbury was as far away from Hartford as you could get and still be in the state. Before the interstate was built in the 1960s, a trip to Danbury by the governor was a rare event as it required a two-hour road trip over U.S. Route 6, which meandered through the many towns that were scattered along the route. So, a visit by the governor was a big deal!

Connecticut Governor John Lodge, John W. Leahy, and "nephew," Jack Stetson. *Clarence F. Korker*

As Uncle John's sidekick, the first governor I met was Governor John Lodge with his wife, Francesca. The governor and his party were welcomed to Danbury with a luncheon held by the Rotary Club in town. Following the luncheon, the attendees, which always included the current mayor and every local politician and captain of industry, caravanned their way to the Fair.

A pair of state police cruisers, fore and aft of the governor's car, would escort the governor onto the grounds and up to the front of the Administration Building. A uniformed security detail would open the doors of the big black Cadillac limousine transporting the governor and the official couple would emerge, accompanied by the State Adjutant General in his uniform, to be greeted by John and Irv...and me.

I was eight years old and, boy, was I impressed! John had the humorous habit of introducing me as his uncle and as the next manager of the Fair. So this was how I was introduced to Governor Lodge when I met him in 1952.

Mrs. Richmond of Greenwich (All Palomino Sheriff's Posse), producer Mike Todd, "Buck" Steele (Steele's Frontier Days), Hazel Steele, John Leahy, actress Elizabeth Taylor, C. Irving Jarvis, Jack Stetson.

Gladys' son and Jack's father,
John Stetson, working at the beer stand.

The governor and his wife were strikingly good-looking people. They both had been movie actors before Lodge got into politics. After that, I thought that all governors and their wives must look like movie stars. During the thirty years in which I was to meet governors, I found that wasn't necessarily true.

Uncle John made it possible for my parents to earn some extra money during Fair Week by having them run the beer stand in the center of the grounds. They paid the Fair the same rent that anyone else would have paid, but this was a big bonus for my family. They hired friends to work the beer stand and, although the effort required long and hard hours, it was well worth it.

Every night after the Fair closed, the entire beer stand staff went to dinner courtesy of my father. After dinner, a few returned to the grounds to stock and ice down the beer for the next day. When I was old enough to drive, they hired me and my good friend, Conrad Kasack, to load and ice the bins each night. We were paid $10 for each night's work and made a good week's pay.

Naturally, that's how Connie and I acquired a taste for beer. We always slept soundly after each night's work!

In 1966, my father died of a heart attack suddenly at the age of 45. After that, I partnered with my mother in running the beer stand during Fair Week. I did this for three years until I changed jobs in 1969 to become the full-time Assistant Superintendent of Rentals for the Fair, which I'll share more about later.

Uncle John's idea of a vacation was to visit fairs, festivals, parades, circuses, zoos, trade conventions and trade shows—exhibitions of any kind. It was fun, but always work-related. There was no wasting away beachside for him.

When I was a little kid, I accompanied him each year to Member's Day at the Bronx Zoo ("I'm a member of the zoo," he'd chortle), as well as going to the Ringling Brothers Barnum and Bailey Circus at Madison Square Garden, The Rhinebeck Fair in Rhinebeck, New York, the Middletown Fair in Middletown, New York, the Eastern States Exhibition (otherwise known as "The Big E") in Springfield, Massachusetts, the 1964 New York World's Fair, the Macy's Thanksgiving Day Parade, and any other local celebration of any kind. Often Gladys and later my wife, Carol, would join us.

While our early trips were all within two hours of home, when I turned 12 I started to travel with John on more far-flung adventures.

One evening, shortly after the run of the 1956 Fair, Uncle John called the house. He wanted to know if I was interested in accompanying him to the Texas State Fair in Dallas. I had never been farther than Philadelphia to visit my maternal grandparents on the train, and that was long before. Excited to travel by airplane for the first time, I eagerly helped Mom pack my bag. These were the days when plane flights were still special. So I needed a suit in which to travel and attend the special luncheon for visiting fair managers to be held at the fair. In preparation, Uncle John dispatched me to Feinson's Men's Store on Main Street to get suited up.

An American Airlines DC-7 awaited us at Idlewild (now Kennedy) Airport, ready to whisk us to Dallas in only about five hours. Uncle John seemed almost as excited as I was as he explained the different happenings and noises that were part of taking off and landing. These were the days when passengers were served a choice of three entrees for the meal, which was served on a white table cloth with a cloth napkin and real silverware. I was impressed!

John was the Regional Director for the International Association of Fairs and Expositions (IAFE). That position, plus his amiable and humorous personality, made him popular among the nation's fair managers. Also, he was the only one to own his own fair, and that made him unique. All the other county and state fairs were run by boards under the jurisdiction of the various departments of agriculture. He could make his own decisions and spend his own money without supervision (except for some mild discussions with Gladys and Fred Fearn,[71] which, of course, he always won).

The other fair managers in the association were employees of a board and had to convince them to go along with their plans and ideas. That status gave him special entrée at the fairs we visited. We simply phoned the fair president to announce our arrival and a car was sent to our hotel to pick us up. A visit to the fair manager's office left us with a fistful of tickets for any of the attractions being offered. Also, we were invited to special luncheons for the governor or IAFE events.

It was at one of these events that I was to meet Leo Carrillo who played to comedic sidekick Pancho to Duncan Renaldo's Cisco Kid. I watched them on TV every Saturday morning and was delighted to meet him in person and get his autograph. I was surprised to find out his Mexican accent was for TV only. In real life, he spoke clear unaccented English.

The trip proved to be a successful adventure for me and it must have been for Uncle John as well. When July 1957 rolled around, I was invited to travel

---

71 Fred G. Fearn was the manager of the fuel companies and also had overall responsibility to make sure expenses did not exceed income.

with him again. This time to the famous Calgary Stampede in Alberta, Canada. This trip involved a total of ten hours in the air with Air Canada, three hours to Toronto where we changed planes and then seven more hours aboard a Lockheed Super G Constellation four-engine turbo-prop to Calgary.

While there, we stayed at an old railroad hotel, The Palliser, in the center of the city. From our floor, we could see the fireworks emanating from the fairgrounds not far away. Especially impressive were ground pieces[72] that were portraits of Queen Elizabeth and President Eisenhower framed by both national flags in all the colors that fireworks could present.

The Calgary Stampede was a large fair in the traditional sense, but it was known for its nationally sanctioned series of rodeos that culminated in a national rodeo championship on the final day. The whole city celebrated the opening day of the Stampede with a big parade through the center of the city.

Stampede manager Maurice Hartnett invited us to view the parade from the balcony of City Hall. He promised to send a car to pick us up at the Palliser. That morning, a liveried driver found us waiting in the lobby and led us outside to a waiting limousine. Along with the limo was a six-motorcycle police escort. We got into the limo and off we went—three motorcycles in front, three in the rear, and all with blue lights flashing and sirens screaming. What a reception!

As visitors traveled from Goldtown to the New Amsterdam Village, they were greeted by The Pirate's Den, 1972. *R.W. Mannion*

Mr. Hartnett greeted us at City Hall and he and John had a great laugh over his practical joke. I never forgot that event or the fact that Uncle John was so well known and popular all the way out in Calgary. After that welcome, I hate to admit, the parade was anti-climactic.

February 1959 rolled around and I accompanied Uncle John to the Florida State Fair and the Gasparilla celebration in Tampa. Gasparilla Day celebrated the legendary invasion of Tampa by pirate José Gaspar. It involves an "invasion" of

72 A firework display that is low to the ground and consists of varying colors to make specific images.

the city by a fully rigged pirate ship and rowdy participation by 700 of Tampa's most prominent citizens, all decked out in pirate regalia. This thrilling spectacle has over a 100-year history that continues to this day.

The Greater Tampa Showmen's Association was holding a luncheon to which we were invited while we were at the fair. Roy Rogers and Dale Evans were performing there and they were also present at the luncheon, seated directly opposite John and me at the table. The conversation at the table was about their raising of Santa Gertrudis cattle on their ranch. I wasn't involved in the discussion, but was in awe of being at the same table with the "King of the Cowboys" and the "Queen of the West."

Each Thanksgiving Eve, John attended the National Showmen's Association gala held at The Commodore Hotel at Grand Central Terminal in New York City. This association consisted of carnival owners (showmen), carnival game owners and concession owners. The dinner was a get-together where they could talk about the successes of the last season and parade their wives around in all the diamond jewelry and other finery that they owned. It was a formal affair and Uncle John was always invited to sit on the dais.

Jack and Carol Stetson, 1973.

The first year we were married, John invited Carol and me to accompany him. He sent me to Feinson's Men's Store to buy suitable clothing and Carol hand-sewed a striking baby blue formal gown for the occasion. We took the train from Brewster, checked in at the hotel, and were soon immersed in a crowd of some of the most interesting people I was to ever know. All summer, these people lived on carnival lots and fairgrounds across the country. They were competitive, tough and streetwise, but also maintained an insular brotherhood that worked as a kind of mutual aid society, which was especially important since "carnies" were never held in high esteem by the outside world.

The dinner opened with the band playing "There's No Business Like Show Business" as all the officers and honored guests paraded to the head table in their tuxedos, while Uncle John introduced me to the other guests as "the next manager of my Fair."

After dinner, we all headed to our hotel rooms for a good night's rest since tomorrow, Thanksgiving, would be another big day. Francis Messmore, a supplier of floats for the parade and a friend of John's, arranged for us to have seats on the bleachers right in front of Macy's store where we would watch the parade in sometimes frigid weather. Next, it was back to the train to make our way back to Danbury. We would arrive at the Leahy house just in time for the big turkey dinner Gladys had prepared. The next day, Gladys would drive John to Harmon Station, New York, to board the train known as "The Commodore Vanderbilt" for an overnight trip to Chicago to attend the Fair convention. (I joined him in later years.) Those few days of peace and quiet allowed Gladys some much needed recuperation time.

Years earlier, when I was only 15, I had another kind of introduction to the fair business await me at the joint annual convention of the International Association of Fairs and Exhibitions and the National Association of Parks, Pools and Beaches. This convention was held at the Sherman House Hotel in Chicago starting the Saturday following Thanksgiving. This trip was another of John's working vacations and I was invited to go with him.

The ritual was this: The very day after the grueling thirty-six hours that contained the Showmen's dinner, Macy's parade and Thanksgiving dinner, Gladys drove us to Harmon, New York, where we boarded the train to Chicago, an overnight trip. I was treated to another adventure that included meals in the elegant dining car and sleeping in a tiny stateroom. Climbing into the top bunk, I was soon lulled to sleep by the constant click-clack rhythm of the train traversing the tracks.

Disembarking in the morning in Chicago, we arrived at the hustling convention being held at the Sherman House. The convention consisted of a schedule of luncheons, dinners, presentations by fair managers and a unique trade show.

The trade show was held in a vast display hall. One could see the newest developments in kiddie rides, fair food products and carnival games all in action. At the show, I got to play Whack-A-Mole before it was introduced to the general public.

A feature of the meeting each year was a film presentation by Uncle John about the Danbury Fair. Irv Jarvis had grown up with Fred M. Carley, who became a documentary film maker. His subjects were mostly of hunting and fishing exhibitions sponsored by sporting goods manufacturers, but every year he volunteered his services and shot film of the Fair. It was raw, unedited film, but that's how John wanted it. He used it to study each year's Fair to find ways to improve it. But, he also took the reels that were shot and his 16-mm projector to Chicago to show off his unique Fair to all who were interested. (These were the films from which Irv Jarvis, Jr., and I created a 90-minute video about the Fair in 1998.)

21

JOHN'S KEY TO SUCCESS

The ultimate key to success for any entrepreneur is surrounding himself with the right people to help achieve his goals and manage the details of the business. Having been successful in his previous ventures, John realized he knew nothing about the nuts-and-bolts of the Fair business and needed someone to fill that gap. He had dispensed with most of the pre-war "old guard" as he reworked the Fair to encompass his own ideas and start with a clean slate. Yet he found in Charles Irving Jarvis the person he was looking for.

C. Irving Jarvis

"Irv" had grown up in the amusement park business at the nearby Kenosia Park. The park was previously been the site of a small resort anchored by the Hotel Kenmere. The Hotel offered rental boats for rowing or fishing on Lake

Kenosia, tree-shaded grounds for picnics, concerts and other outings. It was operated by Irv's father, William Jarvis, and his brother-in-law, Leo Lesieur. Another uncle, Charles Morey, visiting from Massachusetts, thought it was a likely spot for a carousel. Being an entrepreneur himself, Charles partnered with William and installed a merry-go-round. Eventually, they built a roller coaster and added other rides and games as well.

The two purchased the Hotel Kenmere and established a theatre on the ground floor. It was in this atmosphere that Irv grew up as an active participant in the business. As a young teenager, he became the projectionist for the silent movies of the time. Since he had learned to play the piano, he would start the projector and then race downstairs to play the music that accompanied the films. When intermission came, he sold popcorn.

One night Irv and his family narrowly escaped death when the Hotel burned down. He launched himself out the second-story window of his bedroom onto a lower roof, and then jumped to the ground from there. Unfortunately, the hotel was never rebuilt.

After that, he spent a few years employed by Atlantic City's famed Steel Pier, owned by the prominent promoter, George Hamid. Here, Irv broadened his experience in the amusement business.

During the war years, he was employed by the Barden Bearing Co., helping to manufacture the famous Norden bombsight. This invention was critical in improving the accuracy of the bombs dropped by the military during the war.

Irv was also an artist, designer and inventor. He invented a Bingo machine that was built by Bethel cabinetmaker, Joe Vaghi. He also fashioned a "sound car" out of his Uncle Charlie Morey's old Pierce Arrow. He mounted speakers on the roof and set up a mobile sound studio in the back seat, complete with turntable, amplifier and microphone. He and the car were available for hire for ball games, outdoor festivities and political rallies. He was hired by the City of Danbury for several years to play Christmas music to the shoppers on Main Street. A driver would slowly cruise Main Street, while Irv played Christmas records from the back of the car. Irv was also instrumental in designing the one-fifth mile midget racetrack at the Danbury Fair.[73]

Both William Jarvis and Charles Morey had become superintendents at the neighboring Danbury Fair. Irv was their assistant and became familiar with the inner workings of the annual event.

Given his background and experience, Irv was just the guy John had been looking for. He joined John's circus and assumed the many duties of Assistant

---

73 Background on Irv Jarvis from telephone interview with his son, C. Irving Jarvis, Jr., August 2015.

General Manager. These included event promoter, racing pit steward, announcer, negotiator, show booker, concession rental manager, public relations person, artist, researcher and overall detail man.

On top of his official duties, Irv had to deal with every spontaneous idea that jumped out of John Leahy's mind—and that wasn't always easy. The two spent many hours together, especially during the winter months, hammering out the details for each year's Fair.

Being native Danburians, they also knew all of the town characters and their stories, and enjoyed reminiscing about old times. Raucous laughter could often be heard emanating from Irv's office as the two of them reminisced about their experience with Danbury's past.

Irv Jarvis and John Leahy making plans in New Amsterdam Village.
Danbury Museum and Historical Society

22

MY CAREER BEGINS

My working career for the Leahy organization began when I was in seventh grade. *The Danbury News-Times* classified section sponsored a category of free ads for students to find summer work. I submitted an ad looking for lawns to mow. I received one reply from a friend of my father. I mowed his lawn only once before Fred Fearn called me and asked if I would be "office boy" at Leahy's. The pay was $1 an hour, cash.

Weekday mornings, I punched out little steel plates with new customer addresses on an Addressograph machine used to print invoices. I also made address changes. Then on Saturday mornings, I swabbed what seemed like an acre of appliance showroom floor with a string mop and a bucket. After that, I dusted all the appliances on the floor and touched up any blemishes I found. Each week I went home with $20 in my pocket—quite a haul for a twelve-year-old in 1957. I continued this routine all through high school.

When I turned 16 and could drive, I spent summers working on the maintenance crew at the Fair. Uncle John loved having young people around and hired a bunch of us every summer. We painted, mowed and hauled the large statues into place after refurbishment during the winter. We also put the antique carriages and wagons into place, hung dozens of signs, picked up hundreds of hay bales gleaned from the grassy parking lots and stacked them into the barns, re-shingled roofs, and helped out the carpenters, electrician and the plumber.

As high school and college-age boys, we also delighted in performing all sorts of mischief. This job was as much like summer camp as it was employment and we enjoyed it to the fullest.

Uncle John's vehicle of choice was a black 1936 Ford coupe. (This replaced the 1935 Ford coupe he'd had and loved earlier.) He customized it a bit to conform to his idea of a fairgrounds work vehicle. He had a pair of steel garage door handles bolted to the roof of each side. These were intended to be handholds for workers who stood on the wide running boards as John drove them to job sites on the grounds. He had the trunk lid wired open and had a steel-lidded toolbox built that protruded from the trunk and extended out over the rear bumper. The freestanding headlights were perfect for wrapping one's legs around while sitting on the wide front fenders. With this configuration, he could carry six of us on the outside and one in the passenger seat as he drove us to our on-site destination.

A group of us would be working together when John would drive up in his old coupe and holler, "Hop on, boys!" We would eagerly all run to one side of the car and hop on, overloading the suspension, and making it lean severely to one side. John would lean out the window and holler, "No, no! Some of you get on the other side!" With that, we'd all vacate the one side and load up on the other. This would leave John to exit the vehicle and individually assign each of us to a designated spot on the car. We'd laugh all the way to the next location—and we were certain John enjoyed it too, judging by the slight smile on his face.

Instead of having any kind of lock-and-key system, John seemed to have a different padlock on every gate and building door. He carried dozens of keys on a leather thong attached to a pine stick measuring about 2 inches by 12 inches. On the stick he had printed in big, bold pencil, "Please Return to John W. Leahy" followed by his address and phone number. He would unlock a building, walk in, lay the key ring down somewhere, get engaged with whatever, and then walk out of the building, snapping the padlock shut, and leaving the keys where he had set them, going on his way. This happened often. Soon, he'd ask, "Boys, have you seen my keys?"

It wasn't too long before John would offer a $5 reward to anyone who could find his keys, but most often he would retrace his steps and find them himself.

Soon, of course, the $5 was too much to ignore. One of us would find this massive key ring on the seat of the coupe when John was otherwise engaged and make it vanish until the reward was offered. Of course, we couldn't pull this off more than twice a summer—John wasn't stupid!

One morning, as we gathered around the time clock waiting for work assignments, John drove up and said, "My car smells really bad. We have to clean it out and find out why." He routinely would stop and pick up any papers, rocks, sand or other debris that had found its way onto his pristine fairgrounds and toss it into the open trunk. Eventually, this needed to be emptied.

Every summer, John purchased about fifty white Pekin ducklings to raise in the New England village pond and have swim around the paddleboat during Fair Week. Apparently, some predator had killed a duck that year, but was scared away and left the carcass by the side of the pond. John had discovered it, picked it up and flung it into the car trunk, then forgot about it. We discovered the rotting carcass as we emptied out the refuse-filled trunk. John and a few of us headed to the Fair dump to dispose of it. When we got there, someone proposed that we have a funeral for the poor duck. We found an old shoe box, lined it with some papers, laid out the duck, and graced it with some nearby wildflowers. We then formed a procession, humming the Funeral March as John led the way. Coming to a likely location, the shovel bearer excavated a grave and Ducky was laid to rest. We all had a good laugh and went back to work.

The summer was filled with hot, humid days of hard work and more shenanigans involving everyone on the grounds. Raucous laughter often filled the air. And we got paid too!

After graduating from Danbury High School, I became a college boy. John wanted me to become an accountant, so I began that course of study at the University of Rhode Island. I soon found I hated accounting with all its debits and credits and innumerable fudge factors. After three semesters of accounting gloom, I left college and came home.

I joined the fairgrounds maintenance crew, married my high school sweetheart, Carol Farwell, then got a "real" job as a propane gas installer and appliance serviceman. Carol and I lived in an apartment in a big old Victorian on South Street for $85 a month plus electricity, which ran about $15 in those days. Carol worked as a bank teller at Connecticut National Bank.

After a year on South Street, Uncle John called and asked if we would be "caretakers" of the fairgrounds. The idea was that we would live at the fairgrounds in a portion of the farmhouse that had been part of the Backus Farm. John had divided it in half. Part of it was used as the public school art display in the Fair. The other part had been modernized to make it livable. The former maintenance foreman who had lived there with his large family had moved to greener pastures and it was vacant.

The caretaker's house at the edge of the Airport (Blue) Parking Lot was the residence of Jack and Carol from 1965-1970. Circa 1966.

It was pretty grimy from heavy use over several years, so John had it cleaned and painted. Then we moved in. Shortly after, John had a rather large sign installed under our bedroom window that said, "Jack and Carol Stetson, Caretakers."

In return for free rent and electricity, there were duties involved. I was to feed and water the animals every day before and after work. Saturday mornings, I went to Meeker's Hardware and loaded up on feed and hay for the next week. "The animals" was a shorthand way of describing a flock of about eighty black Karakul sheep, a dozen llamas, and a few goats. In the summer, the flock of fifty baby ducklings was added. They were to grow into the noisily quacking fleet that followed the boat in the New England Village pond.

In the winter, this became a little more difficult. As the Fair's water lines were buried only a few inches below the surface, they were drained each year after the Fair so they wouldn't freeze and burst. Also, the electricity was shut off to most of the grounds to lessen the possibility of fire.

Those circumstances added to the job. I carried a large flashlight to the animal corrals as well as a large wrecking bar, which I used to break up the ice that had frozen in the water containers overnight. Water came from an old hand pump. I would fill five-gallon buckets and transfer the water to the animals' water buckets, which were made from the domes of hundred-pound gas cylinders that had been scrapped. They made perfect water tanks.

Of course, when it snowed, the roads had to be plowed in order to reach the animals who resided at the opposite end of the grounds from our house. Also, John wanted all the roads plowed for fire truck access, in case of fire.

Plowing was another adventure. The vehicle was a red and grey 1946 Model 2N Ford tractor mounted with a snow plow. The plow angle could only be changed by the operator hopping off the machine, pulling a pin in the mount, and adjusting the plow by hand. The tractor had no protective cab so the operator was continuously exposed to the snow.

Plowing with the cold wind blowing snow in one's face for several hours required some perseverance. Afterward, I would stagger home red-faced from the cold with snow covering my hat and icicles hanging from my glasses.

All of this was in addition to my regular 54-hour per week gas installer job. We did this for five years, lived off my earnings, and banked what would have been rent money in addition to Carol's teller pay. By 1969, we were able to buy a building lot and get a mortgage to build our house. We were grateful for the opportunity to save some money. Uncle John didn't give away money, but he always provided ways to earn more. Of course, he always got the better part of the deal.

23

CHANGING WITH THE TIMES

John W. Leahy had his own vision of what would make the Fair special to the changing tastes of fairgoers. Though the Fair was started as an agricultural exhibition, the farming way of life was rapidly disappearing from the Connecticut scene.

Increasingly, the Fair was attracting a largely suburban audience whose only exposure to raising crops was in mowing the lawn and its experience in animal husbandry was raising the family dog or cat. While wishing to retain the traditional agricultural bent of the Fair, he knew that changes and additions had to be made in order to keep the attendance growing.

One of the first things he did was to remove the coin boxes for the bathrooms and do away with the parking fees. Both of these freed up congestion that caused traffic in the immediate area (both by foot and vehicle) and made attending the Fair that much more enjoyable. Adding to this, Uncle John also repainted the

The giant Indian Chief greets visitors in Goldtown.

buildings, paved the walkways, added benches, and provided garbage cans, seemingly every few feet, to reduce the litter around the fairgrounds.

Without any visible long-range plan, Uncle John instituted new attractions upon impulse even as early as his first Fair in 1946. He began a collection of larger-than-life fairy tale characters that eventually numbered in the hundreds. These were purchased at fire-sale prices from defunct amusement parks and department store displays.

John Leahy and eight of his giant reindeer.

He introduced himself to Francis Messmore of Messmore and Damon, a New York City builder of displays and parade floats. Mr. Messmore maintained a warehouse full of used display items and had his finger on the locations of others around the country that were due to be scrapped. He soon developed a feel for the types of items that would appeal to Mr. Leahy and alerted him to their availability.

These almost always required extensive refurbishing to be displayed at the Fair. The result was the creation of a staff of talented carpenters, artists and painters who worked on them during the spring and summer each year to get them ready for display at the Fair in the fall. As this talented staff grew, many of the figures were created in the fairgrounds workshops.

Leahy also discovered manufacturers of giant display figures. He enjoyed showing off the figures he purchased by placing them atop buildings and at prominent locations around the grounds. One of the earliest was a nearly 100-foot long display of Santa Claus and his eight gigantic reindeer as they ascended skyward. The popularity of TV cowboy Gene Autry's record "Rudolph the Red-Nosed Reindeer" caused the addition of that warning light-equipped creature as well. The whole display included 30-foot long candy canes.

Other notable giants were a 20-foot high cow and a bull mounted on the roofs of the cattle barns, 30-foot high figures depicting Paul Bunyan, Uncle Sam, Farmer John, an Indian chief, Rip Van Winkle, a cowboy that remarkably resembled actor Clark Gable, and a giant Dutch boy whose countenance closely resembled Mad Magazine's character, Mortimer Snerd.

Big Duke, the world's largest ox, watches Wally Smith with his Brown Swiss calves.

An oversized replica of the entire Budweiser eight-horse Clydesdale Hitch, complete with beer wagon, drivers and Dalmatian mascot "Bud," was mounted on the roof of one of the horse barns. The Hitch honored the Fair with its presence every couple of years as they participated in the Daily Street Parade and the Grandstand Show. Of course, their beer was the exclusive brand on the fairgrounds.

Many more of these figures, large and regular-size, graced every nook-and-cranny of the grounds. Part of the fun of a Fair visit for the kids was the discovery of these at every turn.

THE BIG TOP

Although it wasn't a Leahy innovation, a long-standing tradition was the Big Top, the virtual centerpiece of the fairgrounds. The original Big Top, which had been destroyed by fire, had five center poles. When it was rebuilt in 1941, it was done with only four center poles. The fifth became the flagpole in the racetrack infield.

The tent was not anchored to the ground as most tents are. Fronted by the new Administration Building, an oval-shaped wooden structure was built to enclose a courtyard. The wooden

A Connecticut Yankee at the Danbury Fair.
Danbury Museum and Historical Society

building housed commercial exhibit spaces, and the canvas tent was attached to the roof of the building, covering the courtyard.

Irv Jarvis described it thusly for a press release:

THE BIG TOP

Twenty-seven thousand square feet of billowing white, eight-ounce circus twill in ten sections, each section lashes to the other to form an area 120 feet wide by 225 feet long.

To keep this in the air, with four steel center poles embedded in concrete, the tent is raised on these by large blocks and pulleys and then tied down by steel cable. Centers reach 45 feet in the air, the 10 half poles are inserted equally under the Big Top a quarter of the way down the canvas and 24 quarter poles another quarter of the way down, making a total of 34 poles plus 4 center poles. This bracing holds 7,500 lbs. high up in the air when the sun is shining, but triple its weight when wet.

Seven thousand seven hundred fifty-five feet of rope is sewn into the sections to firmly support this weight and also to give the Big Top its shape. The ends of the canvas on this huge oval have extended lash lines that tie onto a traffic cable strongly bolted on top of the building section. This cable starts the incline of the Big Top from 14 feet to 45 feet, or "pitch" of the tent as it is called.

The four centers are attached to 2-foot bail rings. These bail rings in turn are on the block and tackle that raises and lowers the tent.

After the Big Top is in position and firmly lashed down, and has been tested by the Fire Marshall and state police, the business of display and exhibit takes place. Here, under the Big Top can be found the granges, fruit, flowers and vegetable displays along with what is known as "The Big Top Circle" of local and national exhibitors.

Big Top superintendent Arlene Yaple and Governor John Dempsey.

Those displays were all organized and supervised by the Big Top superintendent, Arlene Yaple. A farmer herself from New Milford, she was also the *News-Times'* New Milford reporter. She truly knew the farmers and news from up-county. Each year, she came up with a theme for the granges to follow, assigned them table spaces, and kept abreast of the condition of the displays, as some of the produce would begin to spoil over the course of the Fair.

The appearance of the Big Top canvas over the fairgrounds was a sure indication that the Fair would soon open. It was also a source of intense worry for John Leahy. In light of the history of the tent, he feared that fire or storms would rip the canvas from its lashings and destroy it. The top went up two weeks before the Fair opened in order to set up all the tables, the stage and the bleachers, and give all the exhibitors plenty of time to prepare.

For many decades during the Fair, Al Stone would come on duty in the evening to help close the area, assist the exhibitors out the door, then begin a lonely night-long vigil for errant cigarettes or other dangers that might befall the tent. After the Fair closed each year, exhibitors were strongly encouraged to vacate as soon as possible as the tent came down about three days later.

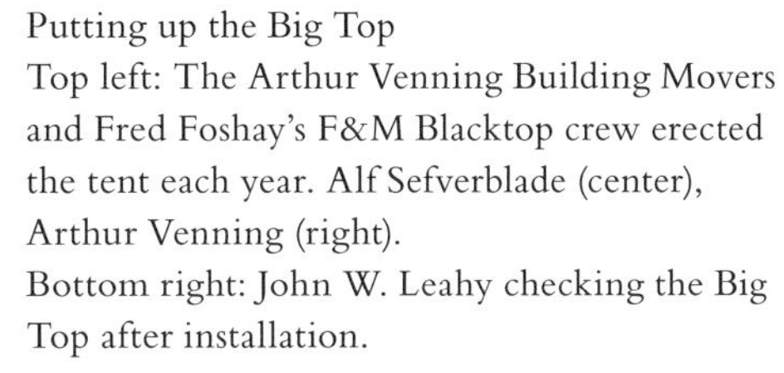

Putting up the Big Top
Top left: The Arthur Venning Building Movers and Fred Foshay's F&M Blacktop crew erected the tent each year. Alf Sefverblade (center), Arthur Venning (right).
Bottom right: John W. Leahy checking the Big Top after installation.

The weather forecast for Wednesday, October 3, 1979, called for the possibility of strong winds and heavy rain. John scoured the local hardware stores for all the 3/8" or ½" rope they had in stock. The idea was to reinforce the tent by lashing more ropes to the anchor cable and running them over the tent at intervals, tying them off on the other side.

Someone had to walk the seams of the sections to carry the ropes over the top. The seams were the strongest places, especially as this tent was nearing the end of its useful life. That someone turned out to be me since I was the lightest person around—I weighed probably 140 lbs at the time. So over the tent I scooted repeatedly, dragging the ropes behind.

Seeing the Big Top floor from the opening at one of the center poles, 45 feet straight down, gave one pause, especially when only an eighth of an inch of canvas provided one's support.

The sky darkened early in the afternoon and the winds and rain came up as promised. Everyone at the Fair was warned to try and take shelter via public address announcements. The Big Top was evacuated by all but our maintenance personnel. The wind whipped and the quarter poles actually jumped up and down off the ground, but never broke free of their tie-down stakes. The additional ropes did their job and the storm blew through quickly on its way to Hartford. Damage on the fairgrounds was limited to a few downed signs, torn tents and a muddy parking lot.

However, as the storm reached the New England Air Museum at Bradley International Airport outside of Hartford, it generated a tornado with winds of over 200 mph, destroying most of the museum's planes and structures. The first building of the rebuilt museum opened almost two years later, in 1981. The Fair had dodged a major disaster.

The Parade

As was mentioned earlier, John Leahy loved a parade. He never missed a local one, especially Danbury's annual Memorial Day parade. He was always invited to view that spectacle by Marion Green from the front porch of her Green Funeral Home as it passed by on Main Street.

Naturally, he thought nothing would be more fun than to stage his own parade at the Danbury Fair. This happened not just on opening day, as was the case with the circus parade when that spectacle arrived in a town, but every day the Fair was open starting with opening day of the 1946 Fair. And what a spectacle it was!

Every afternoon, Harold Kohler, John's cousin, who was normally employed as a carpenter refurbishing the Fair's figures, became the official Parade Marshal. It was his job to line up the dozens of parade units that were to march throughout the

Harold Kohler, carpenter and Parade Marshal,
Charles Wildman, electrician. *George B. Keeley*

John Leahy and Irv Jarvis,
decked out as ringmasters.

grounds. This was especially difficult when the daily crowds swelled to over 60,000 attendees, jammed cheek-by-jowl among the Fair attractions. At precisely 2:30 every afternoon, the crowds were parted to make way for the start of the Danbury Fair Street Parade.

The parade consisted of virtually everything on the grounds that had wheels or hooves that could be driven or hitched-up to proceed along the route. It was led by John Leahy himself, accompanied by Irv Jarvis. The pair dressed in traditional ringmaster's costumes, complete with top hats, scarlet coats, white jodhpurs[74] and black boots.

In the early years, they walked ahead of the parade, carrying fancy walking canes. They delighted in waving to the crowd, especially to

Cinderella's pumpkin coach follows her slipper
in the Danbury Fair street parade, 1950s.

74 White riding pants that were cut very full over the hips, tapering at the knee to become tight-fitting from the knee to the ankle.

all the local people they recognized along the route. However, in later years, they rode in a horse-drawn hansom cab with a uniformed driver behind a pair of handsomely-matched black horses.

Following behind them were dozens of floats drawn by gaily painted pickup trucks that had been retired from Leahy's propane business, antique farm wagons and elegant restored coaches of many varieties. Among these was an authentic Wells Fargo stage coach, a replica Conestoga wagon and a large heavily-barred animal wagon containing an enraged King Kong gorilla. Then there was the elegant jewel-encrusted Queen's Coach, the purple pumpkin coach of Cinderella, a circus bandwagon carrying the Zembruski Polka Band, and a float featuring Princess Goldenrod with her band of Penobscot Indians from Raymond, Maine. Several "musical" wagons transported the assorted bands that played around the grounds.

Victor Zembruski Polka Band. *Danbury Museum and Historical Society*

"Princess Goldenrod" (Dorothy Ranco), Penobscot Indians, receives trophy from John Leahy, 1952. *Clarence F. Korker*

Always the final unit, and the most exciting to the crowds, was the famous eight-horse Budweiser Clydesdale Hitch. In years when the Bud Hitch was not available, its place was filled by the Hallamore Hitch of Clydesdale horses from Massachusetts.

Interspersed among these were many local marching bands and fife-and-drum corps, including the Danbury High School Band, the Danbury Drum Corps, the Grassy Plain Drum Corps from Bethel, the New Fairfield Sparklers, the Nathan

Hale Ancient Fife and Drum Corps of Coventry, and the Troubadors of Bridgeport. On occasion, others also were included.

As if this weren't enough, on Wednesday, traditionally Governor's Day, the parade included every cow, bull or calf that could safely be paraded, attended by the farmer who brought them.

Irv Jarvis, Danbury Fair Queen Vicki Mills and John Leahy, 1953. *Clarence F. Korker*

This entire ensemble wound its way through the crowded midways each day, pressing the fairgoers four deep against the concession stands. It then found its way onto the racetrack and in front of the grandstand, where each band would perform to the delight of the thousands who gathered to witness this spectacle.

Jack Stetson, Danbury Fair Queen Ellen Gerety and SSgt Ron Klutts, 1974.

At that point, Uncle John and Irv would approach the microphone and greet the crowd. Also, the Queen of the Fair would say a few words and be escorted to her seat by a member of the United States Marine Corps.

As amazing as it seems, the only casualties over all those years was the occasional foot that was stepped on by the giant hoof of a one-ton Clydesdale or Belgian horse.

During my boyhood, Uncle John always invited me to ride with him and Irv in the parade. That was fun, but the parade eventually provided me with a memorable public embarrassment.

Sometime around 1955, the Fair had an exhibitor who was a commercial airline pilot from nearby Ridgefield. As a side venture, he was promoting the idea of families having a burro as a family pet. They were reputed to be docile, cute, fuzzy, rideable and perhaps easier and cheaper than a pony to keep. He prevailed upon Uncle John to find a boy to ride one in the parade in hopes of garnering some free publicity.

John had the answer immediately and soon I was introduced to a grey-brown burro turned out with a red leather saddle, stirrups and reins. I was already decked-out in my Hopalong Cassidy cowboy outfit (white hat, black shirt and pants, and a pair of matched pearl-handled pistols). After a couple of brief lessons in burro-manship, I was sent off for my mounted debut in the parade—alone.

I fell into line behind the Grassy Plain Drum Corps and proceeded to follow the line of march, waving to all the onlookers as I had been instructed. I sure was cute! The parade proceeded into the crowded grandstand arena. As was customary, the parade was halted each time a musical unit was front-and-center so that it could perform a number for the audience.

At its turn, the drum corps halted, did a right-face, was introduced and began to play. At a respectful 30 feet or so, I gave the Stop command to Burro (or so I had named him en route). Pulling lightly on the reins, I intoned, "Whoa."

Now mules, donkeys and burros, being of the same general family, are known to be strong, intelligent…and stubborn! They often can't be persuaded to go. Burro went fine—but his stubborn streak kicked in when it was time to stop. As Burro and I continued to approach the band, I pulled the reins a little harder and more emphatically ordered, "Whoa!"

As the band played a spirited John Phillips Sousa march, Burro and I invaded its ranks. I continued my entreaties by rearing back with tightened reins and now panicked commands, "*Whoa, Burro. Whoa!*" Burro continued to march on, now traveling between the third and fourth ranks of the band, and the audience was in stitches!

At this point, I had no choice but to let the animal continue its way through the middle of the band, out the other side, and on its merry way back to the barn. Little did I know that my first solo adventure in a parade would turn out to be a comedy act.

Apparently, the effort to popularize the burro as a family pet didn't succeed—and there *certainly* is no burro in my backyard.

Mlle. Gabrielle and Silhouette, the Dancing Stallion, performed in 1952. *Clarence. F. Korker*

The Grandstand Show

The end of the parade was a natural segue into the grandstand show, which had existed for years already. Most large fairs enlisted big name stars at considerable cost, which appealed to only a narrow segment of the audience. John Leahy's idea was to scour the country for unusual and exciting acts that would entertain everyone. As mentioned earlier, John often spent his "off" hours visiting smaller fairs, circuses and the annual trade convention in Chicago each year looking for something different. He combined these new acts with the best of what the Fair had presented in years past to create something the crowds were sure to enjoy.

The 5,000-seat main grandstand and flanking bleachers provided seating for nearly 10,000 people, so the action had to be big in order for all to see. The Hunt Brothers Circus provided their biggest and most daring acts during the first few fairs of Leahy's management. These included trick riders on horseback, liberty horse acts,[75] a high-diving horse that jumped some 50-odd feet into a small water-filled pool, the human cannonball, and of course the elephants.

The high-diving horse. *Frank Baisley*

Since John had flooded the racetrack for boat races, the circus performed on the infield surrounded by this moat. A bridge had been constructed near the fourth turn in order to allow entrance to the infield. When the Hunt Brothers' herd of elephants was to be guided across to the infield led by Dolly, they refused to cross

75 Riders control their horses through voice and hand gestures while doing acrobatics and tricks. No harnesses or reins are employed.

Dolly, followed by Blanche, leads the way over the moat to the midget race track, circa 1948.

John Leahy's bridge. Perhaps they sensed something about John's construction method that they didn't like.[76] After a while, circus owner Charles Hunt cajoled them to cross with food and other entreaties, much to the delight of the crowd.

Other years saw stock car and midget auto racing, but the noise drowned out the rest of the Fair's activities during Fair week and so was discontinued. Favorite acts among the fairgoers included demonstrations of the driving skills of the Budweiser Clydesdales, the swaypole act of the Nerveless Nocks, and the marvelous precision riding exhibition of the matched chestnut horses and smartly uniformed riders of the Royal Canadian Mounted Police Musical Ride.

The Nerveless Nocks performed 80 feet above the ground on flexible poles. Members of the group would climb the poles and set them to swaying rhythmically. As the tops of the poles crossed paths, the daredevils would switch places from pole to pole, performing acrobatics while they did so. This act held the audience in rapt attention and suspense.

The Royal Canadian Mounted Police Musical Ride was created to honor the tradition of the mounted policemen who served in the rugged terrain and desolate small towns across Canada.

76 John had a tendency to tell his carpenters, "Don't build it too good, guys. It's only temporary. Just build it good enough."

The Ride consists of thirty-two horses and riders that perform a thrilling and delightful precision riding routine accompanied by martial music.

The thirty-six members of the Ride are selected from over 900 applicants. Each serves a three-year term with the Ride. About ⅓ are rotated out each year. The horses are sixteen to seventeen hands high and ¾ to ⅞ thoroughbred stock. The sight and sound of the lance-carrying scarlet-coated Mounties charging about the grassy infield was one of the main memories of fairgoers. They visited the Danbury Fair several times over the years.

To augment the opening of the Fair's Goldtown Western Village in the mid-1950s, John wanted to create a cowboy-themed extravaganza, reminiscent of Buffalo Bill Cody's Wild West Show, to perform in front the grandstand. Along came Buck Steele's Wild West Show.

Buck's show consisted of a lot of horses, cowboys and Indians.

Ivan the Great, the Human Cannonball.

The Nerveless Nocks, swaypole artists and headliners for The Greatest Show on Earth. *Danbury Museum and Historical Society*

John Carlson's Shetland Pony Hitches race for the entertainment of crowds at the grandstand.

John W, Leahy salutes The Royal Canadian Mounted Police Musical Ride. *Danbury Museum and Historical Society*

Another form of ostrich racing, with carts rather than riders. Just as much fun, though! *Danbury Museum and Historical Society*

Sol Solomon performing one of his high-diving acts. *Frank Baisley*

Following a loosely composed script authored by Buck with the help of Irv Jarvis, the show was largely a melodrama involving hordes of mounted Indians attacking a wagon train. This effort, of course, was defended by equal hordes of rifle-firing cowboys. The highlight of the show was the burning of the cover of one of the Conestoga wagons as it raced across the field, spewing black smoke and big flames induced by a liberal application of kerosene. This show was both memorable and laughable, and isn't likely to be repeated at any fair anytime soon.

Another in the list of memorable shows was Gene Holter's Wild Animal Show, in the late '60s. Holter carried with him a menagerie of lions, tigers, cheetahs, leopards, ostriches, camels, horses and elephants "straight from the movie lots of Hollywood."

The highlights of the show involved volunteers from the audience. Some were invited to wrestle with the tigers, a dangerous stunt to be sure (who insures this stuff?). That created plenty of "oohs" and "aahs." Next were camel races. The participants were helped up on the backs of a pair of special "racing camels" and set off on their merry way running a complete lap around the racetrack, followed by

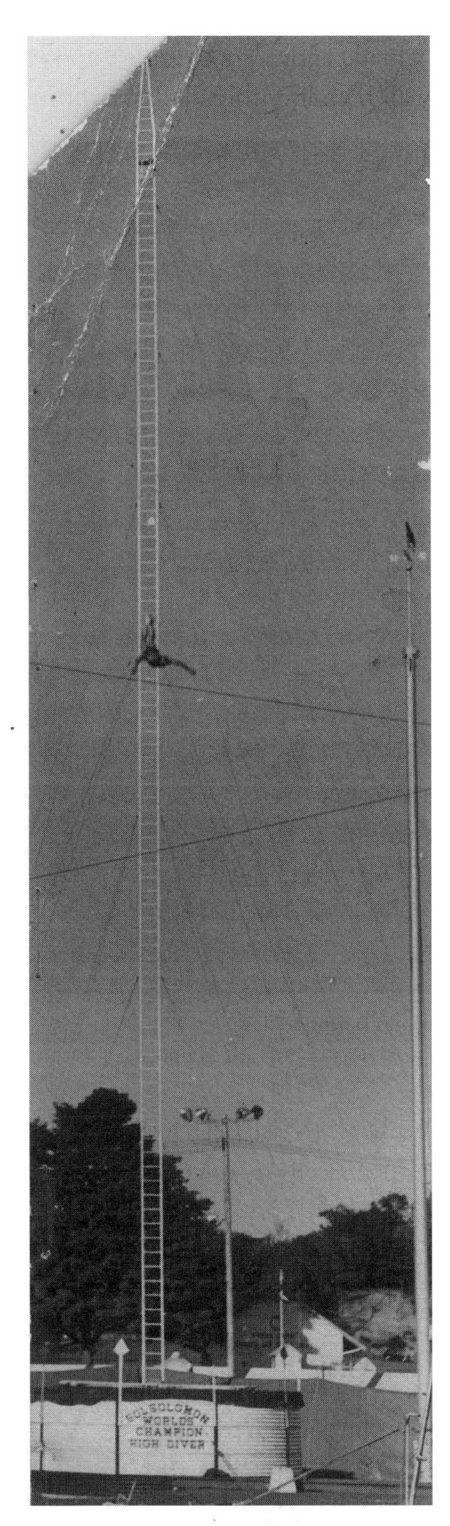

Sol Solomon, World Champion High Diver, jumps from over 100 feet into a tank of water only 6 feet deep, 1949.

horse-mounted outriders.

The ostrich races were by far the strangest. The riders, each equipped with a broom, were set astride the giant birds equipped with ordinary ostrich saddles. Being untamed and unpredictable, the birds were steered by holding the broom up to the eye opposite the direction one wanted to go, since the birds would only go in the direction they could see. For instance, if a driver wanted to go left, he had to hold the broom alongside the ostrich's right eye to block the bird's sight in that direction.

Ostriches are very fast, capable of speeds up to 45 mph. The race also required horse-mounted "safety riders" to gallop at breakneck speed alongside the birds. The main purpose of the outriders was to keep the birds in the confines of the racetrack and to stop them at the end of the race. They were ignorant as to the meaning of a checkered flag.

The final and funniest act was to see how many people could sit on the back of one of the huge elephants. Again, volunteers from the audience came down to participate. They would be helped upon the elephant by two of Holter's staff. Each person was boosted up on the elephant by sliding forward from the tail end. This, of course, became more hilarious as the numbers mounted, all accompanied by the verbal encouragement of the announcer, Gene Holter. Twelve to thirteen people usually succeeded in crowding on, their antics delighting the crowd. The show left everyone laughing and a good time was had by all.

One evening, the week prior to the opening of the Fair, my wife, Carol, and I decided to take a walk through the grounds and check in with the Holter gang that had settled in that afternoon. The late September chill required that we wear heavy sweatshirts for comfort. We came upon the group finishing supper around a picnic table, being entertained by one of the animal handlers cavorting on the ground with a young cheetah.

Now big pussycats have a lot in common with little pussycats. They have the same pert ears, intelligent eyes, cute fuzzy faces and long tails. We were introduced to the cat, who seemed very frisky and playful, and I was invited to join in the fun. After a few tentative swats, I was soon knocked to the ground for a little rough-and-tumble with the fun-loving animal. Weighing about 60 lbs and composed of solid muscle, I soon realized I had more than met my match. One of the animal trainers thankfully separated us after a short while.

Upon regaining my footing and my composure, I found my sweatshirt had been shredded and a bit of blood dripped from my unprotected hands. The other thing big cats have in common with little cats is a set of lethal claws. The cheetah had used them liberally while toying with me and I was very glad the chill required such a heavy shirt!

The Thrill Show, Auto Daredevils, Hell Drivers... All names for the same type of show involving precision driving, smoking tires, roaring engines and smashing of junk cars. These shows dated back to the invention of the automobile and probably sprung from the excitement of early auto racing. The Danbury Fair had featured these for many years. Automotive thrill show names that had graced the Fair racetrack since before World War II included Ward Beam, "Lucky" Teter, "Irish" Horan, Jack Kochman, "King" Kovaz, Joie Chitwood, and finally, Bob Connally's Hurricane Hell Drivers.

The excitement was real, most of the stunts were dangerous, and everyone in the audience was thrilled to see professional drivers performing stunts that we all wish we could try in our own cars. The stunts included side-by-side ramp jumping, crisscrossing on the narrow speedway at high speed, crashing through

Not to be tried at home, auto polo provided thrills, spills and chills as teams raced to push a ball around the racetrack. *Frank Baisley*

flaming wooden barriers, and balancing a car on two wheels while circling the track. A favorite act included the mayhem of the famed "T-bone crash" involving intentionally launching one junk car over a ramp and smashing it into another placed at right angles to the oncoming vehicle. The final act involved the firing of a car from a simulated cannon, which hopefully landed on a distant receiving ramp amidst the earsplitting explosions of fireworks.

All of this was accompanied by the patter of an overexcited announcer with comedic breaks offered by the track clown while the clean-up crew prepared the track for the next event. These clowns were variously named Crash or Sparkplug or Retread.

The clown typically arrived in a little beat-up, garishly-painted, badly running Crosley. He would then climb out of the car through the window and would be greeted by the announcer, "Good afternoon, Crash. That's a fine looking automobile you've got there."

"Thanks, but it don't work so good."

"Why? What's the matter with it?"

"Well, the carburetor don't carb."

"Wow, the carburetor doesn't carb? That's too bad. What else is wrong with it?"

"Well, the generator don't gen."

"Oh boy! Is there anything else?

"Seems there's a problem with the piston."

"So, what's the problem with the piston?"

"Well, the piston, she don't work either!"

That kind of simple, silly comedy punctuated the whole show and laughing at it even while rolling one's eyes was irresistible.

As the young official of the Fair, I got to spend some time with a few of the show performers. I soon grew friendly with stunt driver Bob Connally who purchased the King Kovaz show and eventually became my brother-in-law. Learning of my desire to

Motorcycle crashes through flames, 1972.

try some stunts in a car, he invited me to try a few things on the racetrack infield. The stunt we practiced was known as the "Reverse Spin." It involved driving forward at about 30 mph, then spinning the car 180° and coming out of the spin going backward.

The sequence of performing this stunt went like this: accelerate to speed, pop the transmission into neutral and cut the steering wheel hard to the left while braking, which caused the car to spin. After spinning the car, pop the transmission into reverse and accelerate—hopefully in the same line of travel on which the stunt was initiated. After several tries at this, I soon gained a lot of respect for the drivers who did this on a tight racetrack every performance.

I was also privileged to ride as a passenger during a show while racing through some of the precision driving stunts. The most exciting was the balancing of the car on two wheels while circling the entire racetrack. My driver was "Bumps" Willard. He drove the driver-side wheels up a ramp at slow speed which launched the car sideways. Bumps then calmly toyed with the steering wheel until the car achieved balance, all the while smoking a cigarette. Just another day at the office!

Meanwhile, my body was pressed tightly to the passenger side door, my ear inches from the pavement, which I could view in close-up fashion through the open window. About two-thirds of the way around the track, the front A-frame that supported the suspension collapsed, catching Bumps' cigarette-holding right hand in the steering wheel as it spun wildly under the weight of the car. We came crashing down, right-side-up thankfully and were able to limp back to the front of the grandstand to be greeted by the announcer and a cheering crowd. So much for my adventure as a Hell Driver!

Jack Kochman was one of Lucky Teter's successors in the world of auto thrill driving. His Hell Drivers changed their act every year, adding new and more spectacular stunts.

24

A TOUR THROUGH THE VILLAGES

Always the innovator, Leahy searched for ways to vary the attractions at the Fair to broaden its appeal and attract larger audiences. As the attendance continued to grow, he also needed to use more of the grounds to accommodate the numbers.

New England Village - 1950

Trips to New York City via the New York, New Haven and Hartford Railroad to attend the circus and various trade shows landed John in Grand Central Terminal. On several occasions, he was able to view a life-size diorama called "Main Street New England" sponsored by the railroad and mounted in one of the porticos high above the main floor. The village was comprised of a half dozen building facades that represented what a small nineteenth century New England

Just a quiet New England Village on a beautiful afternoon.
Danbury Museum and Historical Society

town center might look like. The centerpiece was a typical white columned church with its spire reaching for the heavens. The church was 60 feet tall. The other storefronts included a print shop, a general store, a blacksmith shop and a couple of others.

Realizing this was a temporary display, John contacted the railroad and arranged to purchase the village when it had outlived its tenure at Grand Central. It was deconstructed and shipped to the Danbury fairgrounds.

An underused area on the western edge of the grounds became the site for the "New England Village," which opened for the Fair in 1950. The site was flanked by a large red horse barn complete with a second floor haymow. The barn had been used to house work horses and maintenance equipment for years. As the village was comprised of only false fronts, the Fair work crew built buildings behind the fronts to serve as exhibits or concessions. Additional buildings were added, bringing the total number to eleven.

Three square ponds were dug out in front of the street. Culverts leading to nearby Kenosia Swamp enabled water to flow in and out of the ponds, maintaining its level. To complete the scene, John added a flock of live white swans and schools of goldfish. A few weeks before each Fair, a flock of perhaps fifty white Pekin ducks joined the swans and added to the avian cacophony.

Eventually, the ponds were joined as one and two islands were created. A replica of a Mississippi steamboat, the "Miss Priscilla," gave rides around the pond, pushing the flock of ducks out of the way, quacking loudly as they made way for the intrusive boat. One should note that historical accuracy was often compromised in favor of the spectacular at the Fair villages.

The Fair had become the owner of the former Harvey Backus farm, which sat at the corner of Kenosia and Backus Avenues, a site formerly known as "Backus Corners." The farm's large barn was moved by the Arthur Venning Building Movers company and installed opposite the old red barn on the south end of the pond. This became the abode of the Home Arts display for many years, which was eventually displaced by an animated display of the Dr. Dolittle story. When interest waned in that story, the barn was renamed "The Art Barn" and housed exhibits by local artists and craftspeople.

The Backus farmhouse was dismantled into sections. The smaller portion became the one-room schoolhouse in the Village and displayed the art work of students from the Danbury public schools, supervised by Rose Geary, the head of the schools' art department at the time. The remaining portion was sited near the Airport Gate and became the caretaker's residence.

During his weekend tours of the countryside, the purpose of which was to visit Fair stockholders and persuade them to sell him their shares, John noticed many abandoned horse-drawn wagons and old farming implements that had outlived their usefulness. Soon the word was out and owners of this type of equipment would offer their derelict items to him. The result was a collection of dozens of wagons and farm tools of all types. These were cleaned, restored and placed around the New England Village becoming an outdoor museum, which also added to the charming ambiance of the area.

This collection also included an authentic horse-drawn trolley car, an early railroad coach once owned by the Boston and Providence Railroad, an authentic stage coach built by the famed Brewster Company, and a Conestoga wagon. The crowning piece of the collection was the acquisition of the "Daniel Nason."

When the New York, New Haven and Hartford Railroad decided to reduce their inventory of antiquated equipment, it included an 1858 steam engine, named the Daniel Nason, and its tender among its items for sale.

Not many people could purchase, store and maintain a full-size locomotive, but John Leahy could and proved a ready buyer. Thus, the venerable

John W. Leahy admires his new toy train at delivery, early 1950s.

old piece of railroad history found its way to the Danbury fairgrounds in 1951.

The Nason was designed and built under the supervision of George S. Griggs in the Boston and Providence Railroad yards. It is the last remaining example of a particular type of locomotive popular in the pre-Civil War era known as the "Dutch Wagon." This design had the steam cylinders mounted inside the locomotive frame. Originally a wood-burner, it was later adapted to burn coal. It had 54-inch drive wheels and 16 by 20-inch steam cylinders. It was beautifully painted black with red and green accents and gold-leaf scrollwork.

It was an outstanding addition to the New England Village where it was displayed for several years. Later, it was relocated twice more on the grounds before eventually finding its way to the National Transportation Museum in St. Louis, Missouri, where it is now the oldest steam locomotive in their collection.

Goldtown Western Village – 1955

One of the Fair's streets was known as Church Row for decades. Dedicated church members of various local denominations prepared and sold ethnic food items to raise money for church projects. This required tremendous effort on the part of the volunteers as it required many hours of labor and a large number of people. As more women began to go to work full-time, it became increasingly difficult to continue these homegrown operations.

Fair management recognized this and opted to replace Church Row with something currently in fashion. There was now a television in nearly every home and the adventures of movie cowboys were now transferred to the small screen. Gene Autry, Hopalong Cassidy, the Cisco Kid, the Lone Ranger, Roy Rogers, *Gunsmoke*, *Maverick*, *Have Gun, Will Travel* and countless others filled the evenings and Saturday mornings. The newly popular *World of Disney* ran a miniseries starring Fess Parker as Davy Crockett. Kids all over the country ran around with coonskin caps perched on their heads and the series' theme song became a hit.

The Dance Hall at the Golden Nugget Saloon.

The natural thing to do was to capitalize on this theme. The Church Row buildings were white with green trim, the predominant color of all the Fair's buildings. These buildings were very old, had the typical flat fronts that old stores used for signage, and were lined up in a row along the street. It didn't take much imagination to envision a typical cowboy-era Western town.

The buildings soon got a coat of grey-brown paint. Some were sided with slabs of wood with the bark intact. Covered front porches and boardwalks were added to give that movie-set western town look. The signs were painted with western-style lettering declaring them to be the Golden Nugget Saloon, the Barbershop, the Sheriff's Office and Jail, the Blacksmith Shop, the Bank, and the Goldtown Hotel.

Behind the jail was the Boothill Cemetery complete with wooden gravestones, their epitaphs stolen directly from Tombstone, Arizona. Standing ominously at the rear of the cemetery was a gallows from which hung a recently executed desperado, swinging in the breeze, having died with his boots on. Such macabre fun!

To add to the flavor of the gold rush, dozens of sand-filled burlap bags were scattered about, each labeled "GOLD!" Almost every flat surface was covered with old horseshoes and cow skulls. Real cow skulls!

In his quest for some authenticity, John put in an order for several dozen cow heads from Abie Novak's slaughterhouse on the eastern outskirts of town. These came delivered in barrels, not only with the pre-requisite horns, but also complete with their eyes, meat, skin and hair.

They needed to look old and sun-scorched as if they were collected from the arid desert of the West. A little research revealed that if the heads were boiled in a solution of water and sulfuric acid, all the soft tissue would disappear into smoke, leaving only the bleached bones, ready to be nailed up all over Goldtown. The New England Village farm implements collection had several giant cast-iron cauldrons that were commonly used to scald recently slaughtered pigs. These were mounted above large propane-fired burners, the prescribed solution was added, and the cow heads were tossed in to boil furiously until the meat was gone. All of this produced a stinking, sickening slurry. Over the course of several days this activity continued, to be tended by a succession of employees who became sickened by observing, and

Panning for gold in Goldtown. Note the cow skulls used for decoration.

The original farmhouse and caretaker's residence, summer home of John and Gladys Leahy, later became the Goldtown Hotel.

smelling, the process. At last it was over and Leahy had his collection of authentic skulls to augment the decoration of Goldtown.

A three-story tall farmhouse stood at the end of the Goldtown main street. This was one of the oldest buildings on the grounds and once served as the caretaker's home. It was christened the "Goldtown Hotel." Even though it was used as the residence of John and Gladys Leahy during the summer months and Fair Week, it became part of the Western lore. Large signs were posted declaring it a hostelry. The second story windows were opened to display a group of comely blond-wigged, heavily made-up female mannequins watching the street scene. A couple were even mounted on the porch roof, rocking to and fro in electrically powered rocking chairs. Hotel, indeed!

The Ona-May Dancers at the Goldtown Music Hall, 1957.
Clarence F. Korker

A theatre, the Goldtown Music Hall, was constructed seating about 200 people. Dance students from the local Ona-May School of Dance performed the can-can and other old-time dances a couple of times a day. Country singer Stella Donnelly and her band also graced the stage there, as did Red Brigham and the Harmony Hayriders.

A picture of the Old West wasn't complete without a train to be attacked by outlaws and Indians alike—and the Fair already owned a steam engine—the Daniel Nason. Art Venning was enlisted to relocate that enormously heavy machine to Goldtown. A ramp was built, complete with railroad tracks, and the ancient engine was slowly winched up onto a heavy-duty flatbed truck. The ride was about a ¼ mile. Tracks had been laid on a strip of land opposite the Goldtown buildings. The operation was reversed and the engine and its tender were now ensconced in a new scenario.

A real flat car and an authentic caboose were obtained and added to complete the little railroad. The flat car was loaded with old steamer trunks and wooden shipping crates, all appropriately labeled for their western destinations. Adding some flavor, Indian arrows were drilled into the crates, and they were further adorned with painted bullet holes. Kids were delighted to clamber all over the train, and clamber they did!

"Red" Brigham and the Harmony Hayriders had a standing invitation to play at the Fair. A 1962 article mentions he appeared at the Fair for six consecutive years.

A bearded, leather-clad prospector wandered up and down the street accompanied by his donkey. The donkey was laden with gold mining tools—pick, shovel, gold pan and supplies to aid the miner in his search for the yellow bounty.

The bank was robbed several times a day and a gunfight was re-enacted. The Sheriff's posse always saved the day and the bad guys were unmasked and thrown in jail. Goldtown was a success and became one of the most popular annual attractions at the Fair.

Stella Donnelly was another regular performer at the Fair.

New Amsterdam Village – 1960

As the attendance at the Fair continued to increase, the density of people trying to enjoy all the exhibits made it increasingly difficult for all to walk around in comfort, especially on weekend days when crowds could swell to upward of 50,000. Recognizing this, and also desirous of continuing to add to the attractions of the Fair, John needed a new idea.

The Dutch Windmill Restaurant, 1960s.

Mr. Gillotti with his Scotch Highland cows, 1969.

The appeal of the traditional agricultural fair was slowly changing as the western Connecticut landscape became more urban and suburban oriented. Increasingly, the people attending the Fair didn't know a cow from a bull and thought chicken came from a factory, frozen and wrapped in cellophane. With the opening of Interstate 84, throngs now visited from New York City and its environs.

The first Europeans to settle and develop New York City were Dutch immigrants. The list of accomplishments by these people is long and interesting. In combination with that factual history were the legends of the Hudson River Valley to the north, including Washington Irving's stories "The Legend of Sleepy Hollow" and "Rip Van Winkle." John and Irv decided to capitalize on this history and the idea for the New Amsterdam Village was born.

The old half-mile horse racing track had been abandoned since before World War II, but the former backstretch remained a dirt road lined by a number of old horse barns used only for hay storage and assorted junk that had accumulated over the years. This lay to the north side of the currently used one-third mile stock car track. The area lent itself perfectly to further development.

Irv spent a great deal of time during 1958 and 1959 researching the fact and fiction of early New York. He also drew up proposals to convert the old horse barns and surrounding grounds to depict a village that might be reminiscent of the time when the Dutch culture influenced daily life. All that was missing was a person to head up what was to become a major design and construction project that was too large for the existing Fair staff.

Hans, in New Amsterdam Village, 1972.

Mark D. Isselee was a talented, enthusiastic and industrious builder and entrepreneur. He had been a partner in the small petting zoo, amusement park and restaurant establishment on Post Road in South Norwalk known as "Old MacDonald's Farm." Due to a falling-out with his partner, he was available for the job and jumped right in.

Isselee soon had a complement of carpenters and masons on the job. They worked furiously to get the Village ready to open for the 1960 Fair. It resulted in the construction and remodeling of twenty-six buildings of all shapes and sizes. The old dirt racetrack backstretch was paved and became the main street, lined with antique-style street lanterns. Shrubbery was planted and flower gardens created.

The new village included a schoolhouse, a Dutch-style home, a unique twin-spired building intended to be a restaurant (it was never used for this), and the brick Fort New Amsterdam complete with a dungeon, cannons and rampart. Further along the way, one old barn was faced with brick and became City Hall ("Stadt Huys" in Dutch) complete with a courtroom inside. This was accompanied by a small church building, a bakery complete with a beehive oven, and several barns that comprised the farm (the "Bouwerie"). In addition, a couple of life-size windmills were built, their vanes propelled by hidden electric motors.

The whole scenario was brought to life with many interesting additions. The farm included live sheep, cows, ducks, chickens and llamas (Dutch llamas?). A small mountain was built and populated with a herd of goats that spent their days clambering up and down its ramps to the amusement of all.

Peter Reilly, the Fair's resident artist and sculptor, created life-size statues representing Peter Stuyvesant (governor of New Amsterdam), Adam Roelantsen (the first schoolmaster), and a couple of lawbreakers to populate the dungeon and the stocks. There was also a 30-foot long reclining Rip Van Winkle who appeared to be awakening from his long sleep.

Children visiting with Rip Van Winkle, who is wide awake in the 1960s.

And storks! A historic legend of the Netherlands and other northern European countries was brought to America with the Dutch. It was thought to be good luck to have a stork build a nest on one's roof. And of course they delivered babies. So, true to John Leahy's style, Reilly created storks—many storks—to be residents of the giant nests that adorned nearly every roof in the village. A couple were even suspended on wires with the requisite slings bearing babies destined for New Amsterdam.

The flower gardens were planted with hundreds of chrysanthemums in bright fall colors a couple of weeks prior to opening day. The sweet smell of Dutch apple pie drifted in the air, along with the sounds of Dutch folk music being sung by local songstress Mieke Michael.

The New Amsterdam Village was a resounding success. It added a pleasant entertaining and educational venue to the Fair and was increasingly popular as the years went by.

25

LINKED TO AMERICA

Access to the fairgrounds, as was commented on in my grandmother's part of this history, was chiefly accorded by driving on U.S. Routes 6 and 7, both of which passed close by on their way to converging in the town center, and by railroad excursion to the Fair's own siding. As the popularity of the Fair increased, the traffic back-ups on the roads became insufferable. Cars full of excited families became mired in mile-long jams that inched slowly toward the Fair. Not only did this frustration reduce the fun of the Fair, it also limited the number of people who could—or would—attend.

The solution to this was found in the Eisenhower Interstate Highway System. Authorized by Congress in 1956, this system was created not only to create an efficient transportation system for commerce and the public, but also to provide a means by which troops and military support systems could be moved relatively rapidly in the case of defense needs.

Busloads of visitors explore the New Amsterdam Village, 1972. *R.W. Mannion*

The routes for Interstates 84, 95 and 91 were designed and approved in 1957-58. The plan for Interstate 84 passed through Danbury and the exit meant to access U.S. 7 South to Norwalk dumped the traffic off right at the Fair's doorstep at Exit 3. The section from Danbury to Waterbury was opened on December 16, 1961.

By 1973, all of the sections of Interstate 84 were completed and Exit 3 became the main egress to the Fair for traffic from Massachusetts and New York State. The Fair lost a small chunk of its land due to this development and Leahy's propane terminal, which sat in the way of the route, had to be relocated a short distance, but the increase in the car and bus count more than made up for the inconvenience.

The largest increase was in the attendance of bus groups. The price of admission to the Fair had always been discounted to riders on the rail excursions and bus groups, as they handled large numbers of people efficiently and used up fewer parking spaces. As word of the Great Danbury Fair spread, the Fair witnessed convoys of buses from New York City, New Jersey and even eastern Pennsylvania. The buses numbered in the hundreds, especially on each of the Saturdays of Fair Week.

The occupants of these buses were mostly the parishioners of the black churches from the inner cities—and did they know how to enjoy themselves! They ate, they drank, they laughed, they danced, saw all the shows and generally had a real good time. One could actually feel the merriment that existed in their midst. Toward late afternoon, they would wander back to the buses tired, but happy, in time for a sleepy return ride to the city.

The passing years saw the numbers of buses steadily increase—in good part to the development of Interstate 84. Although wonderful for the Fair's attendance, its location at the juncture of three major routes later became the catalyst for its demise.

Various entrepreneurs began to notice the advantageous location of the 143 acres occupied by the Fair. Rumors persisted over the years that the Fair was to be sold and turned into something else.

As reported in the October 7, 1971, edition of *The News-Times*, James Tyler, president of Transworld Bloodstock Agency, proposed a $15 million project that would return the fairgrounds to a horse racing venue. The proposal included building a one-mile track, increasing the grandstand seating capacity, adding 1,800 horse stalls and, of course, pari-mutuel wagering.

Tyler also proposed similar schemes for the towns of Hartford, Newtown and Monroe. Apparently, he approached the mayors and first selectmen of these towns and also the State of Connecticut to gauge interest in the project, but never actually met with John Leahy until he had created significant publicity and generated a large amount of speculation about the future of the Fair.

Finally, he met with Leahy and Fair vice president Fred G. Fearn. They were interested in finding out more details directly from Tyler, mostly to be more informed. They showed no interest in following the proposal further. Enthusiasm for the project never developed at the State levels or in the other towns and Tyler eventually went away.

26

A YEAR OF CHALLENGES — THE FAIR'S CENTENNIAL

The first question on the lips of any interested Danburian in 1969 was what special things would happen at the Fair to celebrate its 100th anniversary. John and Irv had plans that they discussed in sessions at the White Street offices during the early winter months. Before these plans could gain momentum though, Irv had a heart attack that left him homebound.

Typically, the winter season was when Irv, the detail man of the operation, did the booking of the attractions, the negotiations with the carnival owner, and the rental of some 400 concessions. As he was unable to come to the office, the plans and the usual work began to back up. Eventually, he was able to have some work delivered to his home where he tackled it to the extent that he could. Unfortunately, on May 23, 1969, he a second heart attack, which took his life.

John had lost his most able assistant at the Fair and a good longtime friend.

"New" Management

John Leahy had his 74th birthday on June 5. While he was still vigorous and enthusiastic, the task of producing the 1969 Fair was not a one-man job.

Fred G. Fearn had been the business manager of the various other Leahy enterprises that ran in concert with the Fair. He was in charge of Leahy's fuel oil business, the wholesale oil business and waterfront terminal known as The Norwalk Oil Co., as well as the propane gas business.

1969 Centennial Year Management (left to right): Joe Austin, Jr. (Administrative Assistant), Fred G. Fearn (V.P.), John Leahy (President), Jack Stetson (Assistant Superintendent of Rentals), Leroy Paltrowitz (Press Relations). *George. G. Keeley, New Haven Register*

Fred was also the Fair's vice president, but until now that had been mostly an official title. His main duties had involved accounting and tax reporting, which also meant keeping John from spending more on the Fair than it could afford. A full plate, indeed. Yet now he was asked to take on the management of the Fair, including booking acts, hiring personnel and more!

Obviously, more help was needed, and my big break into the field of outdoor show business was about to happen. I had spent summers as a teenager working with the maintenance crews at the fairgrounds. My wife, Carol, and I were currently living on the grounds in the old Backus farmhouse, where I held a second job as caretaker and sheep wrangler. I had also operated the beer concession during the Fair for three seasons. As a boy, I had spent many summer days hanging out with Uncle John as he went from work crew to work crew supervising their progress. All of that added up to my being the Leahy employee with the most experience at the Fair.

I was excited to accept the challenge of assisting Fred with the management of the Fair. Knowing nothing about the inside workings of the Fair, I was led into what was formerly Irv Jarvis' office and instructed on the mechanics of typing up rental contracts for returning concessionaires and filling vacancies for the upcoming Fair. My occupational title was to be "Assistant Superintendent of Rentals." This was funny because there wasn't a Superintendent of Rentals for me assist!

Fred figured that if I ran into trouble with any of the concessionaires, I could say that I needed to check with my boss, the Superintendent, to clear the matter up. I was 24 years old and some of the concessionaires might have been inclined to try old carnival tricks on the new kid in town. Irv Jarvis had a lifetime of experience with these people, knew all the tricks of the trade, and had gained their respect. I, on the other hand, had yet to attain that status.

Little by little, Fred and I worked our way into the action. Fred was signing up all the performers and meeting with veteran department managers. I was having telephone conversations with the concessionaires and meeting with those who stopped in for a visit. I spent days typing up rental contracts as fast as possible on an old Remington manual typewriter to respond quickly to anxious people who were used to receiving their contracts much earlier.

There were about 400 of these contracts to get through, but we soon had a much bigger problem.

The Great Airport Land Grab

The Danbury fairgrounds shared the valley known as the Miry Brook area with the Danbury Municipal Airport. In 1928, a few aviation enthusiasts purchased land that had been used as an airstrip known as "Tucker's Field" and proceeded to develop it further. Federal money flowed for further development as a result of efforts to support World War II and the facility was eventually purchased by the Town of Danbury.

The main runway ended abruptly at a low wire fence bordering Backus Avenue. Directly under that flight path lay the main parking lot of the Danbury Fair. Planes crossed over the Kenosia Swamp and St. Peter's Cemetery as they departed and arrived. It was a favorite pastime of locals to park along the fence on Backus Avenue, right next to the sign warning of "Low Flying Planes," to watch take-offs and landings, especially on clear Saturdays and Sundays.

The Danbury Fair and the Danbury Airport had always coexisted peacefully. The Federal Aviation Administration[77] cooperated by banning direct flyovers of the fairgrounds during Fair Week. Air traffic increased to its annual peak during Fair Week and during this time overflow airplane parking could be seen near the taxiways.

This peaceful coexistence was about to end, however.

For several years during the 1960s, the Danbury Airport Commission, headed by Chairman Paul G. Annable, was pushing for expansion of the airport. Quite a few of the Commission members were operators of local industrial businesses

77 The Federal Aviation Administration is the national aviation authority of the United States, with powers to regulate all aspects of American civil aviation.

Arial view of the fairgrounds, Interstate 84, Route 7 and a portion of the Danbury Airport runways.

that used the airport to connect to their customers. Their enthusiasm seemed to justify the idea that the airport wasn't big enough for future needs.

The latest proposal for expansion was developed by the Boston engineering firm, Metcalf & Eddy, and encompassed the construction of a control tower, a terminal, a 219-acre industrial park and all the amenities that would support a modern jetport. The biggest piece of the project would be a 6,580-foot north/south runway that would obliterate sixty-eight acres of the main parking lot of the Fair. This estimated cost of the project was about $8 million in combined local, state and federal money.

This proposal was of vital concern to the Fair management. Any large entertainment venue is dependent upon access to its site by its patrons. If the Fair were to lose the lot that provided the great bulk of its parking space, then its very existence was endangered. The editorial pages of *The News-Times* were very much in support of the airport expansion. That bias also leaked subtly into the "news" portion of the paper judging by the quantity, placement and wording of many articles describing the elements and progress of the plan. It was time to fight back.

Leahy and Fearn decided they needed an outside consultant to head up the battle. Inquiries among their associates once again resulted in advice from Francis Messmore of the New York display firm. He knew of the perfect person to join the fight for the Fair.

Demetrios A. Sazani was experienced in problem-solving for a number of small amusement parks. He had put together parades and festivals throughout New York State, and knew how to deal with the officials of various towns. He was smart, enthusiastic, loud, bold and brash as only a New Yorker could be. Sazani tended to overwhelm any situation in which he found himself and wouldn't be cowed by persons of title and importance.

"Dee" took up residence in a small converted garage on the fairgrounds and began his attack. He visited the editorial writers of *The News-Times*. He castigated the publisher and editor, and riled up the public who gave support to the Fair with their letters to the section of the paper known as "Opinions of the People."

The Federal Aviation Administration had published a count of airplane "movements" at the Danbury Airport that supposedly documented rapidly accelerating growth in use of the facility. The count was developed using an extrapolation of "activity which occurs within a four day, eight hour per day traffic sample period."[78]

The letter predicted a total count of movements for 1969 to be 151,882. This means nearly thirty-five flights per hour all day, every day, based on twelve hours of daylight. If one were to believe these numbers, a casual perusal of the nearby sky at any time, day or night, should have revealed stacks of planes circling while they waited for landing clearance along with taxiways filled with planes waiting for departure, like what happens at Logan, La Guardia or Kennedy Airports. This was not the case!

To refute these numbers, Dee decided to take our own survey. He hired my brother, David Stetson, and a buddy of his, Alan Lombardi, both Danbury High School students, to document the flights in and out of the airport for several days. They sat in a canvas-topped Jeep CJ, equipped with lined yellow legal pads, binoculars and jugs of water for their survival.

The survey began on July 14 and ran ten hours a day, from 8 a.m. to 6 p.m. for the next seven days, concluding on Sunday, July 20. The weather was seasonably hot and skies were clear most of the week. Their vantage point, directly under the north/south runway, also gave them a clear view of the east/west runway. They proceeded to count the planes landing and taking off and recorded their wing numbers.

They counted 925 take-offs and 906 landings. Of the 585 flights that flew directly over them in the first four days, they found that 272 of these, or 46%, were made by the same few planes practicing touch-and-go landings and take-offs. Extrapolating this weekly "movement" figure, as the FAA did during a sample period, the total number of flights estimated for 1969 would be 95,212 – a long way from the nearly 152,000 used by the FAA in its report.

---

78 Taken from a letter from J.B Komich, Chief, Airports Branch, Boston Area Office, FAA, Sept. 2, 1969.

Our conclusions were, assuming the FAA was honest in its count:

1. Advance notice of the FAA survey was given and every available Danbury-based pilot was asked to fly during that time to build up the count.
2. Nearly half of the flights were by students of the flying schools located at the airport.

Our survey took place during optimal weather conditions under more than 14 hours of daylight in July. If you consider that year-round ideal flying hours are reduced by shortened winter hours, our survey results should reflect peak usage of the airstrip. (Keep in mind, November through February, daylight hours average around ten per day. Those hours are further reduced by bad weather, including snowstorms, days of heavy rain and low ceilings, and that some summer mornings are obscured by fog emanating from the nearby swamp. Certainly those conditions would subtract from the number of flight "movements."

"Help Save the Fair" petition drive under the Big Top, 1969.
R.W. Mannion

Our results hardly indicated a need for a larger airport, but pointed to over-enthusiasm by Airport Commission members.

Sazani continued his fight by actually having Danbury Fair, Inc., offer to purchase up to eight acres of Airport property to augment its growing need for parking space. That action ignited a spark under Chairman Annable as he fired off a letter to us berating the Fair management for going directly to the City Council without advising him first.[79]

Meanwhile, the Danbury public became aroused. Very few Danbury citizens saw the airport as anything other than a rich boys' club. The advantages of having a noise and pollution producing jetport were lost on the people who would end up paying the bill. These same citizens were also considering the building of a new elementary school at Lake Kenosia at the same time, which would lie in the path of these flights. The safety of our children was to be at risk.

79 Letter of August 6, 1969, Annable to Leahy on Airport Commission stationery

The final battle of the airport expansion was to be waged at the centennial edition of the Fair. Sazani proposed a petition campaign to be run at the Fair. A booth was set up directly inside the main entrance to the Big Top where virtually every visitor to the Fair would pass by. The booth was dressed in patriotic bunting and decorated with large red-lettered signs proclaiming "Save the Fair!" and "Stop the Airport!" Thousands of individual petition letters were printed to be signed by Fair patrons and mailed (by us, of course) to Connecticut Governor John Dempsey.[80] The booth was manned by longtime Leahy staff, including Eleanore Blackman, Ethel Pudelko and others, in shifts.

The effort paid off. Governor Dempsey's office was inundated with mailbags full of petitions citing the will of the people. They wanted the Fair—not an airport! The voice of the voters was heard and the entire proposal was soon dropped.

The John W. Leahy Foundation

John worried about the future of the Fair. What would happen when he was gone? Who would run it and for how long? How would the federal and state government's confiscatory estate taxes be paid? Over the years, the Fair corporation had applied for and received non-profit business status. Then as the rules changed, it was reverted back to regular business status. Finally, it was decided that by creating a charitable foundation to own the Fair, the problem of perpetuation would be solved.

The John W. Leahy Foundation was created on May 11, 1962. The idea was for John to donate shares of stock to the Foundation over several years until the Fair was owned by the Foundation. The trustees, starting with John, Gladys and Fred Fearn, would appoint a board (with themselves operating as management) to run the Fair.

Profits from the Fair would be distributed to charities throughout the area. After John's death, his trustee position would go to a "corporate fiduciary institution." (That position was eventually held by the Connecticut National Bank.)

Several years passed. John continued to donate shares of Fair stock to the Foundation until it owned nearly half of the Fair. Then a bomb was dropped! It was discovered that ownership of the Fair business by the Foundation was not legal as it was set up. Under sections 4941(d), 4943, 4944 and 4945(d) of the Internal Revenue Code, a corporation could not be operated by a foundation.

That news created a terrible problem for the longevity of the Fair. John was forced to have the Fair stock appraised and buy it back from the Foundation at fair market value. Hundreds of thousands of dollars had to be paid to the Foundation

---

80 John N. Dempsey (1915-1989) served as Connecticut's governor from 1961-1971.

to purchase back something John thought he already owned! It's unimaginable that the attorneys who drew up the Foundation were not aware of this since the code was enacted in 1954. Yet here we were, with no other recourse.

The re-purchase of the stock was accomplished, and the John W. Leahy Foundation still exists today. It is proud to have made donations of hundreds of thousands of dollars over the years, principally to Danbury area charities, including the Danbury Hospital.

The original intent of preserving the Great Danbury State Fair had been thwarted. It was left to operate with no plans for continuity.

27

THE CHANGING OF THE GUARD

The 1969 centennial edition of the Danbury Fair was a success by all measures. A record paying attendance of 303,440 was counted. The weather was good and the new management team of John, Fred and I were up to solving the daily problems that arose. Fred and I certainly received intensive on-the-job training as a result.

Dee Sazani, who stayed with us through the Fair, was my constant sidekick. He was a tremendous help in educating me as to the ways of the carnival. Keeping the carnival games honest was emphasized as he taught me some of the tricks of the trade.

For example, novelty concessions near the gates selling balloons, stuffed animals, hats, buttons, etc., were prone to expansion. They would rent a 10-foot space, but then little by little expand to 20 or 25 feet. It happened all the time. We would demand that the display be pulled back or charge them an exorbitant fee for the excess space. This always worked—until the next day. It was an ongoing game that kept me more amused than upset.

The next several years ran equally successfully. A rainy day here and there would knock down the attendance. The year 1973 included ten days of perfect weather and the attendance rocketed to 354,954 after having fallen off a bit immediately following the centennial. Management continued to make improvements.

Magician Harry Albacker performs for Julie Nixon Eisenhower and other onlookers, 1972.

More free acts were added all around the Fair. Magician Harry Albacker, Jerry Lipko's Chimpanzees and the Lumberjack Skill Show all added to the free entertainment venues.

Music was everywhere. The Big Top Stage had an all-day music program including organist Emil Buzaid, songs from Virginia Wren, a concert from Wendell Cook's Circus Show Band, and square dance demonstrations called by Al Brundage, and eventually his brother Bob. Outdoors, the Zembruski

Competitors demonstrating their strength and skill in the Lumberjack Exhibition.
Photos: Jeffrey Yardis (left), R.W. Mannion (right)

Virginia "Wren" Cassidy, otherwise known in Connecticut as the First Lady of Radio. *R.W. Mannion*

Wendell Cook, Cook's Show Band, 1969. *R.W. Mannion*

Carolyn Chase on the bass with the Triple A Ranch Gang, 1972.

Polka Band performed in the gazebo in the Main Plaza, choral groups sang in Garden Park. Stella Donnelly and Company, Carolyn Chase and the Triple A Ranch Gang provided country music throughout the day.

Internal improvements were always underway. The early 1970s saw us involved in restoring the vast wagon and farm implement collection. The Daniel Nason locomotive was again relocated and protected under a brand new pavilion outside the refurbished Transportation Museum. Dozens of brightly painted picnic tables were constructed for use around the grounds.

By 1973, the City of Danbury laid a new sewer trunk line nearly to our front door. We spent that winter laying 600 feet of 10-inch gravity sewer line from the grandstand restrooms and 1,100 feet of 4-inch pressure line from the Goldtown restrooms to join the city sewer. This solved a major problem in the operation of the Fair, especially during peak demand times.

The agricultural traditions of the Fair were continued. Before World War II, competition among the local farmers for prizes at the Fair encouraged them to bring their animals for display. The monetary prizes were small, but the celebratory ribbons on display were not and they were displayed proudly by the stalls of the champion animals.

After the war, with area farming diminishing, the Fair found itself paying a stipend to the farmers who were still willing to participate. It was expensive for a farmer to move his animals, employ help and feed the livestock while being away from his farm for nine or ten days.

Farm animal displays included several breeds of dairy and beef cattle, barns full of magnificent draft horses, a swine barn exhibiting tons of grunting porkers, and the Poultry Show with cages stacked four-high to show some of the strangest looking chickens on earth. A sheep barn displayed breeds of wooly bleaters and varieties of playful goats were on display. Norman Husted supervised the sheep display and obliged fairgoers with a sheep-shearing demonstration each day at the Blue Ribbon Stadium.

One farming tradition that continued at the Fair was the Ox Draw contest. Ably presided over by Wayne Allen of Reading, Vermont, with his New England twang ringing over the Blue Ribbon Stadium sound system—it was a perennial favorite among the farmers and a quaint curiosity among the city dwellers.

Horses and oxen traditionally did the heavy work around the farm that consisted of plowing and preparing the fields for planting, drawing the heavy wagonloads of produce at harvest time, and clearing new fields of tree stumps and boulders. It was found that oxen

Oxen pulling a stone boat during a draw, 1966.
Danbury Museum and Historical Society

Tie for first prize in the Oxen Draw, 1948. Left: John Gulford of Conway, Massachusetts. Right: John Ferris, Newtown, Connecticut. Each team pulled 5,835 pounds a total of 28 inches.

(castrated bulls, especially) had more strength and stamina and were calmer than horses. Thus, they became favored over horses for the really heavy work. At the fall fairs, the question of who owned the strongest pair of oxen was decided by the ox draw contest. After tractors took over the farm work, traditionalists raised and trained oxen as a hobby strictly for competition.

Oxen test their strength.

The ox draw pit is a stretch of level ground covered with a thick layer of clay as a drawing surface. The basic contest involved the hitching of a pair of oxen to a stout wooden stone-boat laden with huge concrete blocks. At the drover's signal, the oxen would attempt a "draw." This required the team to pull the load as far as they could to a maximum of six feet. As each weight class of pullers rotated through, the weight on the boat was increased and the process was repeated until the load was too heavy to be moved. Thus, a champion for each weight class was decided and blue ribbons were awarded.

Ox draw competition, 1950. *Clarence F. Korker*

The weight classes for pairs of oxen were: under 2,800 lbs, under 3,200 lbs

Wally Smith of Danbury with his Brown Swiss oxen. *Henry Reichert*

and over 3,200 lbs, sometimes known as the Free For All or Unlimited class. The beauty of the event comes from watching these enormous animals use their bulging muscles to perform these feats of strength. The champion pair in 1964, owned by Wally Smith of Danbury, drew 9,725 lbs the full 6-foot distance. Some of his competitors that year and most years were William Ferris of Newtown, Franklin Streeter of Cummington, Massachusetts, and Olin Maxham of Woodstock, Vermont.

As the 1970s progressed, John Leahy's health had begun to fail. He had always been proud of his work ethic and retirement was something he never considered. His Fair work was his joy and inspiration.

At the beginning of President Lyndon Johnson's "War on Poverty," John had a sign posted in his office that read, "I'm 74 years old and I fight poverty. I work!" That was his social commentary on the welfare society.

During 1973, John experienced a series of minor strokes that began to erode his mental capacity. A major stroke on Christmas Day caused further debilitation. After a few weeks, he was again able to spend some hours each day at the office. However, another stroke in the spring 1974 sent him to Danbury Hospital for an extended period, then to Glen Hill Convalescent Center, where he spent his last year. The 1974 edition of the Danbury Fair was the first since 1946 that lacked his presence. Although Fred, Gladys and I saw it through, a certain emptiness pervaded the atmosphere. It just didn't seem the same without him.

John W. Leahy's death on March 24, 1975, was indeed the end of an era. His life was the epitome of the American Dream. He used his ambition and drive, his winning personality, engaging smile and down-home sense of humor to overcome his meager eighth-grade education and become a success at several businesses: machine shop, fuel oil and kerosene distributor, gas station operator, propane gas distributor, appliance sales and service, and most notably his role as the savior and rebuilder of The Great Danbury State Fair.

He always looked forward to the next new thing and was excited about tomorrow. He enjoyed the challenges of business and in dealing with all the people he knew and who worked with him. He particularly liked parties, dances, weddings, funerals, parades and spectacles of any kind where he could enjoy being with people. A lifelong student of human nature, he often made wry and astute comments about the events playing out around him.

Robert (Bobby) D. Marquis was the concessionaire under the Grandstand, following in the footsteps of his father, Larry. He rented the grounds from us during the off-season to promote corporate outings for companies such as Pepsi-Cola and IBM. He had also ventured into the animatronics field, creating action figures for various clients. He employed a sculptor (his neighbor) in the animatronics venture named Daniel Long.

Jack Stetson, Fred Fearn and Leroy Paltrowitz look on as
John Leahy fires a ceremonial cannon to open the Great Danbury State Fair.

After the death of Mr. Leahy, Marquis raised money from concessionaires and other participants in the Fair to create an appropriate memorial. Long designed and sculpted the resulting memorial, a bas-relief bronze rendering of Mr. Leahy surrounded by various icons of the Fair with this inscription:

> *John W. Leahy, 1895-1975, A Showman whose genuine Love and Understanding of People Endeared Him to the Hearts of Young and Old Alike by His Personal Warmth, Imagination and Drive, He Built the Danbury State Fair Into a Nationally Known Attraction. Placed by Fair Concessionaires, The SNYRA, and Many Friends.*

The sculpture was mounted in a block of granite, placed in the garden near the Arena Gate, and unveiled during the 1976 Fair. It is now on display at the Danbury Museum and Historical Society.

The future of The Great Danbury State Fair was in the hands of the executors and trustees of his estate. This meant that the Connecticut National Bank, Fred Fearn and Gladys were to decide what came next. The fact that the federal government had invalidated the ownership of the Fair by the Foundation left its future in limbo since John had failed to provide any other instructions as to the Fair's continuance.

Several months went by until Fred announced to me that the executors had named him to be the president of the Fair, that I would be vice president and secretary, and Mrs. Leahy would continue on as treasurer.

The John W. Leahy Monument remains on display at the Danbury Museum and Historical Society, 43 Main Street, Danbury, Connecticut.

28

THE BUSY "OFF" SEASON

For many years, John refused to rent out the fairgrounds to anyone else during the off-season. However, that didn't stop us from putting on our own events during the off-season. John always loved to experiment, and while some ventures were more successful than others, they were always worth giving a try.

The Stock Car Races

In spite of all the planning and popularity, the possibility of a stretch of bad weather during the Fair could ruin its financial success. Luckily, rainy days in the fall were scarce. In fact, the sunny Indian Summer days became known locally as "John Leahy weather."

The success of the midget races after the war helped guarantee cash flow to supplement the gamble that was Fair Week. However, as previously touched upon, the dwindling attendance at the midget races and the very modest success of the operettas and boat races led John and Irv to try the increasingly popular sport of stock car racing.

SNYRA had been successful at a couple of locations in nearby New York State, but had trouble finding a permanent location. They seemed to be well organized, had a set of unique specifications for the race cars, and had developed their own rules for racing. Uncle John and Irv were impressed and arranged with SNYRA for a summer-long Saturday night program for 1952.

The agreement was based on a 60/40 split of the gate receipts with the club. As the crowds grew, this turned into one of the most lucrative arrangements that any racing organization in the area had. It afforded good prizes to the winners and a small stipend to even the least successful of the racing teams.

The Danbury Fair Speedways offered a one-third mile dirt track, the permanent 5,500-seat grandstand, adequate lighting and sound system, and experience at handling the gate. The venue was also probably the most attractive racing venue around with a mowed grassy infield, a red-striped whitewashed block wall lining the front stretch, and a very tall center flagpole from which a large American flag waved in the gentle summer breeze.

The Leahy treatment of presenting a racing program was unique. Over the years, he had acquired three 60-inch anti-aircraft lights, which were powered by their own generators. Their filaments were carbon rods that were fed slowly toward an electrode, creating a very bright continuous spark, much like an electric arc welder. The light was reflected by large convex mirrors that were 60 inches in diameter. These were aimed into the sky where their oscillating beams lit the heavens. They could be seen for miles, attracting attention to the fact that the races were on—just in case one couldn't hear them.

The Danbury Fair Racearena under cloudy skies, 1958. *SNYRA*

A spotlight shone upon the giant American flag that hung from the center pole of the infield.

The flagpole had once been the fifth pole of the Big Top before it burned down. At the end of intermission each night, a Pinkerton color guard conducted a flag-lowering ceremony while the national anthem was played. In 1976, in celebration of the bicentennial of the United States, race driver Evie Pierce sang the anthem over the sound system one night, enthralling the crowd with his deep baritone voice.

Without having to check, I am certain we were the only stock car track in the country that had bubbles. At intermission, recordings of public domain music were played (John wasn't paying fees to ASCAP[81] or BMI[82]), and a dozen bubble machines, mounted in the grandstand rafters, came alive causing thousands of soap bubbles to cascade down into the summer night air, delighting everyone.

Saturday night stock car races became the norm in Danbury for the next twenty-nine years, with the exception of a hiatus in 1956 and 1957, caused by a disagreement between the club and Mr. Leahy, who was determined to show them who was boss.

The popularity of the SNYRA stock car races grew rapidly. Local people constituted the list of drivers, pit crews, sponsors and track personnel—a fact that helped build an enthusiastic regular crowd who attended the show no matter what. They weren't deterred by the weather, holidays, weddings or funerals.

Groups of supporters, families and fans alike, tended to sit together. When the admission gates sprang open at 5 p.m., these fans sprinted to the grandstand to secure their accustomed seats. Woe to the casual fan who arrived and chose a seat amidst one of these fan clubs, as he was soon informed of his error and found himself relocated to a less partisan location.

These groups sometimes came to verbal and physical disagreements, often about which driver caused a crash. The Fair's Pinkerton Security personnel would then have to intervene to dampen this enthusiasm. It rarely got out of hand, but the Pinkerton cops began to know where the trouble was likely to arise.

The "dirt" clay racing surface of the track needed a lot of maintenance. It had to be watered when it became too dry—but not too much or it became slippery. Also, the racing "groove," which was the preferred route around the track, became an actual groove in the track's surface. The corners also developed potholes and rough areas due to the sliding autos. The fans who sat near those areas usually went home coated in dust.

Midway through the racing season in 1958, Mr. Leahy decided that it was time to pave the track. Maintenance of the racing surface would be minimized,

---

81 The American Society of Composers, Authors, and Publishers is an American not-for-profit performance-rights organization that protects its members' musical copyrights by monitoring public performances of their music, whether via a broadcast or live performance, and compensating them accordingly.

82 Broadcast Music Inc. is another performing rights organization in the U.S., similar to ASCAP.

the races would be faster, and the fans, especially those seated in the first and fourth turns, would be a lot more comfortable. So, in a spur of the moment decision, John called in Robert Carroll, the paving contractor who had done a lot of work on the grounds. His challenge was to blacktop the entire one-third mile racetrack and have it useable for the next Saturday night. This was a formidable task, but Carroll accomplished it—and it gave way to a new style of racing at Danbury.

Shortly after that, the Danbury Fair Speedways was rechristened the "Danbury Fair Racearena." That was a small change compared to what the drivers and race car mechanics faced. The drivers needed to learn a new style of driving and develop a familiarity with the characteristics of the paved track, while the mechanics had to specify new types of tires, suspensions and engine setups to cope with the unfamiliar conditions.

The boys responded quickly. The races ran more efficiently, the cars went faster, the crowds went home less dusty. As time went on, flathead engine technology became a thing of the past and parts were increasingly difficult to find.

The grandstand, filled to capacity, at the Danbury Fair Racearena. *Danbury Museum and Historical Society*

In 1973, SNYRA changed their rules to allow overhead valve engines up to 311 cubic inches. That marked the beginning of the modified stock car era in Danbury, which meant a whole new level of racing technology became available to the racers, with higher speeds at significantly higher costs.

The average weekly attendance continued to grow. Management added sections of bleachers wherever space could be found and those seats filled as well. We even refurbished the old abandoned first-turn wooden annex and fitted it with seats. By this time, special trophy races saw crowds of nearly 11,000 people in a single evening.

Realizing that the costs of entering a car as a young novice driver were becoming prohibitive, the club talked Fair management into allowing a low-cost Sportsman division. The 1979 race program was expanded to include a spirited Sportsman division program that remained successful until the closing of the track.

The stock car races made local heroes out of a great many drivers over the years. I hesitate to list names for fear of being accused of favoritism. All had their legions of dedicated followers and many were spotlighted as feature race winners. The two names that stand out from the rest, though, were not drivers.

Paul Baker was the morning drive radio announcer for the local radio station WLAD. His voice was familiar to all Danburians, especially to school-age kids who listened intently early on snowy mornings to see if school was canceled. A sports fan, he also announced for the Danbury High School football games and emceed various dinners for prominent locals.

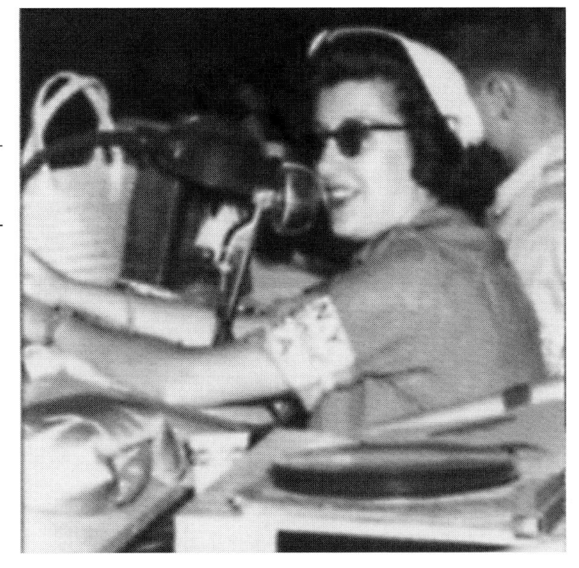

Marge Carpenter, the Fair's racing announcer, 1952. *SNYRA*

According to his story,[83] one Saturday morning he was contacted by a race club official to fill in for our racing announcer, Marge Carpenter, who had come down with laryngitis and couldn't speak. Could he fill in that night as the race announcer?

Paul agreed to give it a try. He knew nothing about the races. He hadn't even attended one. However, Paul used Marge's notes and, aided by the scorers and club officials, he managed to pull it off.

Paul Baker served as the Voice of the SNYRA for 24 years, as well as being the Fair's race announcer after Marge Carpenter. *SNYRA*

The next winter, he got a call from club officials Evie Pierce, Moe Conrad and Bill McCarthy offering him the job permanently. Paul accepted and he stuck with it for the next twenty-four years. Through it all, he became a fan and friend to many of the drivers and officials, as well as being the official voice of the Danbury Fair Racearena.

Many of us can still recite his weekly introduction.

> *Good evening, ladies and gentlemen. On behalf of General Manager John W. Leahy and Assistant General Manager C. Irving Jarvis, and the Southern New York Racing Association, this is Paul Baker bidding you welcome to the stock car racing at the Danbury Fair Racearena, the East's most beautiful racing plant!*

83 Recorded at "The Great Danbury Fair Revue," Palace Theatre, October 12, 2013.

Racearena starters Frank Arnone and Ted Abbott (right). *SNYRA*

Paul presided with calm demeanor over many major and minor crises that happened on the track. Mr. Leahy, customarily seated just yards away from the announcer's booth, was not above begging him to "get the races moving" whenever there was a standstill due to an extended clean-up of a wreck.

Paul's most trying night took place on August 25, 1979. That was the night that Official Starter, Ted Abbott, was killed right in front of everyone.

When the old waterway was filled to create the one-third mile stock car track, the front stretch was left at a grade that was four feet lower than the adjacent infield and the walkway in front of the grandstand. The hope was to keep errant race cars from leaving the track. A block retaining wall lined the face of the infield.

A starter's stand was built on top of this wall as a place for the flagman to store his flags and operate the traffic lights that were placed around the track. The custom was for the flagger to stand directly on the track facing the oncoming race cars. As he waved the green starting flag, he turned and sprinted toward the infield wall and leapt up to safety as the cars roared past. Starters John Coughlin and Hugh "Tiny" Weyer, both large men, added to the thrills each race, week after week, as they accomplished this athletic feat without fail.

As the cars became faster, this practice grew increasingly dangerous. A recommendation by the Connecticut Department of Motor Vehicles, which had jurisdiction over racing at the time, was to build a safer starting stand on the outside of the track. A reinforced steel structure with a platform was constructed to be used instead. It was about six feet high with space for both the starter and his assistant. The fence rail was also increased in height for further protection. The height of the final platform lifted the starter's eyes about 12 feet over the track.

This gave both the starter and his assistant an unprecedented view of the entire racetrack and eliminated the need for them to be on the racing surface. Ironically, it was after this safety improvement was installed that tragedy happened.

Fred Fearn and I were accustomed to working a lot of overtime running the Leahy organization. Our summer weekends included a Saturday workday that started at about 4 p.m. and ended an hour after the races finished. This meant that our workday was usually completed by midnight for twenty Saturdays every year, starting in 1969. This particular summer, 1979, was especially busy as it included six more days for the antique shows, three more days for the craft show, and another day for the auto auction. If those special events took place on a Saturday, we were on duty from 8 a.m. until midnight—a tiring span of sixteen hours.

It was understandable when Fred mentioned to me that he would like to take Saturday, August 25, off to enjoy a weekend with his wife at their vacation spot. However, he was reluctant to do so because of his responsibility at the races. I assured him that I could be the officer-in-charge for that evening. I had been the vice president for four years. All of our people were fully experienced and I worked well with them, so knew I could expect their help in case of a crisis. Thus reassured, Fred left and I was in charge.

The weather was good, the gate opened, the crowds rushed in, Paul Baker gave his usual opening announcement, and the races began. I sat in the President's box—something new for me as I was normally stationed at the gate as supervisor. Alongside me sat Danbury Fire Chief Joseph Bertalowitz, a race fan and frequent guest. In his hand, he held his two-way radio to fire headquarters.

Starter Edmund J. "Ted" Abbott, a seven-year veteran of the Danbury races, dropped the green flag and the race cars were off! On the third lap, the cars were still bunched tightly as they roared out of the fourth turn onto the front straightaway. Suddenly, one car's front wheel made contact with a rear wheel of the car directly in front of it. The first car was thrown through the air. It then rode the top rail of the safety fence, like it was a ramp, and was catapulted even higher when it reached the end of the rail, crashing through the starter's stand.

Assistant starter Frank Arnone instinctively reached to yank Ted out of the way, but the car was too fast and Ted was struck and thrown into the air, landing on the sidewalk in front of the grandstand.

A massive wreck ensued, involving several cars. Arnone was stunned, but unhurt. Smoke and steam drifted over the wreckage, as emergency crews raced to the scene. Ted lay still, face down on the ground. The ambulance and EMTs raced to help along with Dr. Roger Williams, the track physician.

The huge audience was shocked into silence. I ran down to survey the awful scene, then hurried back to Chief Bertalowitz and told him to radio for more ambulances.

Paul Baker spoke calmly to the crowd, assuring them that more help was on its way and that Ted Abbott was in good hands while he was rapidly being transported to Danbury Hospital. As the other ambulances arrived from town, a light summer rain began to fall. Only an uncomfortable murmur came from the crowd. A couple of other drivers were transported to the hospital for treatment, but their injuries weren't serious.

Ted Abbott died that night at the age of 36. The remainder of the program was canceled and the quiet crowd headed for the exit gates. There was nothing left for me to do, but to call Fred and tell him what had transpired. I can only imagine what went through his mind.

Ted loved his job as official starter. He was described as tough, but fair. His decisions weren't always popular, but he stood his ground and called them like he saw them. Needless to say, he was missed by everyone who knew him.

Frank Arnone took over as starter. He probably never climbed into our starter's stand again without reliving that night, but he flagged until the final race in 1981.

The Southern New York Racing Association was not only a racing club, it was a brotherhood. Competing teams would help each other by loaning engines and parts while in the pits, and offering advice to one another while working on their cars during the week.

Even though the track closed more than thirty-five years ago, the club still holds an annual reunion every September. They exhibit their cars, which are all antiques now, while reminiscing about their racing adventures. There is a good showing from the public too, which attends so that they can talk with the drivers and get closer looks at the cars. A lot of the people who come weren't even born when the races ended, but they still enjoy coming to the show.

Although the years are thinning their ranks, they seem to want to continue. Paul Baker often made an appearance until he passed away in September 2014.

The Fireworks

After the financial failures of the operettas and the boat races, John was not willing to use the fairgrounds for other off-season events—except for fireworks.

As mentioned earlier, the history of the Fair included several devastating fires. Fire was a constant worry since there were no fire hydrants to serve the area and the only water available was from the nearby swamp, the pond in New England Village, and a 50,000-gallon water tank that John built specifically for firefighting.

So, it was totally out of character when he decided to present the annual July 4th fireworks shows. John thought that by providing a patriotic concert by the local drum corps while waiting for it to get dark, and combining it with the usual amenities of grandstand seating, food, drink, concessions and restrooms, people would pay to enjoy a celebration of Independence Day.

That's not to say that fire wasn't still a concern. To John, fire prevention was still a high priority. The services of the Miry Brook Volunteer Fire Department were enlisted. The volunteers brought their pumper to patrol the grounds to put out any errant sparks. Fair employees also manned the Fair's own 1954 Ford pumper, which held 500 gallons of water. Others patrolled the area with pickup trucks equipped with extinguishers.

The shows, from 1962 through 1967, were a delightful spectacle. No fires occurred, but the vast majority of the people who enjoyed the show saw it from the surrounding hills and even the Fair parking lots, without paying admission. The number of paying attendees was greatest the first year and trailed off as the years went by. The small profit gained over six years of shows didn't justify the effort and the risk in producing them and they were soon discontinued.

The Charles Ives Centennial Concert

Fred and I had not contemplated other ways to use the grounds beyond the ongoing, very successful stock car racing program on Saturday nights. Eventually, we were approached by Professors Howard Tuvelle and Richard Moryl of the Music Department of what is now Western Connecticut State University. They were heading up a group of local cultural movers-and-shakers who wanted to put on a July 4th concert in 1974 commemorating the centennial of the birth of Danbury composer Charles Ives.

"Charles Ives was perhaps the most innovative composer of the last hundred years—an American with advanced musical ideas, who found himself in an age when this country was dominated by European musical traditions. Then, as well as now, it was difficult for American composers to be heard."[84]

The group's proposal was an ambitious one. They wanted to enlist the services of the world-famous musical conductor Leonard Bernstein to direct the American Symphony Orchestra in performing various Ives' compositions at the Danbury fairgrounds. The point was not only to celebrate Ives' long-overlooked music, but to raise money for an Ives Concert Park to continue his musical legacy in Danbury and to restore his homestead as a museum.

In a most fortunate bit of timing, Tuvelle showed up at the White Street office when Fred, John and Gladys were already having a meeting. Invited to present his idea, Tuvelle met with resistance from John at first, whose experience with "cultural" events was not a happy one. Further discussion finally elicited support from Gladys who would enjoy productions that were a little more "highbrow" than the usual fare at the Fair. John was persuaded and permission was granted for use of the grounds.

Excitedly off and running, Tuvelle began the arrangements. Through the contacts of various members of the committee, Bernstein was approached and he

---

84 Program Notes, Charles Ives Centennial Concert

agreed to present the music. Now the preparations began in earnest.

Money had to be raised to cover the cost of the spectacle. Cultural grants were pursued and obtained from a number of philanthropic sources. For its part, the Fair had to provide the ticket sellers, security, concessions, ushers and other support.

Physical problems were many. The Fair owned no staging equipment and no sound system suitable for orchestra music. Unlike in an acoustically designed concert hall, quality sound had to be distributed along the 1,100-foot front of the grandstand seating area, so an adequate system had to be found. Chairs for the orchestra had to be rented and the chorus needed risers. Adequate temporary lighting had to be installed as well. This was quickly becoming a complicated undertaking.

The City of Bridgeport loaned its portable band shell. Lou Thomsen's Audio Company of Norwalk installed a line of large speakers along the front of the grandstand and arranged for the amplifiers and microphones, and everything else slowly fell into place.

The concert program was to be a presentation by Bernstein, his protégé, director Michael Tilson Thomas, the American Symphony Orchestra, a combined chorus of 200 members of the Greenwich Choral Society and the Western Connecticut State College Chorus accompanied by renowned bass/baritone opera star McHenry Boatright. This concert would be a pinnacle of Danbury's cultural society for years.

Of course, cooperation from Mother Nature was needed. The success of outdoor show business is ultimately at the hands of the weather. Usually, summer means a constant threat from violent pop-up thunderstorms. July 4, 1974, had a

Jack Stetson opens the Charles Ives Centennial Concert, 1974.
Conducted by Leonard Bernstein, assisted by Michael Tilson Thomas.

different problem: bright sunshine, very high humidity and a temperature of 94°.

The orchestra and its support personnel had arrived at the grounds early in the afternoon to review the setup and rehearse the music and the program. They needed time to make adjustments to the sound system and the lighting in preparation for the evening's performance. The participants waited for their turns sitting in the shade of the grandstand, perspiration running freely down their faces. They surely dreaded the prospect of donning their tuxedos for the show.

Consternation with the heat caused a discussion between Bernstein's representatives, committee members and Danbury Fair staff trying to come up with ways to shade the entire area to try and cool it off. However, it was too large to cover and certainly too late in the day to try now.

As the 7 p.m. show time approached, Bernstein's manager began to call for the cancellation of the concert. The heat was melting the wax from the strings of the instruments, while the humidity was making it difficult to keep them in tune. The musicians, used to playing in air-conditioned indoor comfort, were in the throes of near heat exhaustion.

Tuvelle and his committee members pointed to the steady traffic headed toward the fairgrounds down Exit 3 of Interstate 84. They were able to convince the manager that the show couldn't be canceled. Many tickets were sold in advance and everything was ready. The show must go on!

As the sun began to set behind the western façade of the grandstand, a gentle breeze kicked up and the temperature cooled a bit. Mr. Bernstein arrived and the show began as planned. Conductor Michael Tilson Thomas opened the concert and was followed by Leonard Bernstein after the intermission. The music was wonderful, and the crowd of around 7,400 people was delighted. Success was achieved.[85]

THE DANBURY FAIR ARTS AND CRAFTS SHOW

Public interest in craft shows was growing. Perhaps it was a remnant of the hippie movement of the sixties, but the idea seemed to emanate from the living naturally, back-to-the-earth, self-sufficient, anti-establishment philosophy that pervaded the times. There was an increasing number of organized craft shows that exhibited the works of a plethora of very talented craftspeople and artists.

I proposed to Fred Fearn that perhaps we could run a large arts and crafts show to further increase use of the grounds. The success of the Ives concert indicated we could draw patrons to the fairgrounds with more varied types of exhibitions. A couple of field trips to some craft shows convinced him and we hatched a plan.

85 The entire story of the event from start to finish is very well told by Mr. Tuvelle as published in *The News-Times* of July 3, 2005. It is worth the read.

Fred already had plenty on his plate since he was also in charge of the three fuel companies we were operating in concert with the Fair, so he gave me full rein to put the show together. A study of the various craft-oriented publications soon gave me an idea of the schedules of the larger shows in the area, so I needed to pick dates that wouldn't be conflicting. The first Danbury Fair Arts & Crafts Show was scheduled for July 25-27, 1975.

Picking dates was the easy part. We had to get the word out to artists and craftspeople. We advertised in all the popular arts and crafts journals, as well as a wide assortment of newspapers.

Having a separate division for artists (those who worked in painting, sculpture, etc.) was a new wrinkle for this type of show. The line between artists and craftspeople is often blurred. Many craftspeople are artists, many artists do crafts.

We were able to get experienced advice from Rudy Kowalczyk, a silversmith with years of experience promoting large craft shows. He had exhibited at the fall Fair and wasn't afraid our new show would compete with his. He offered his help and we accepted. We also attracted the attention of well-known seascape artist Patricia Yaps, former head of the Connecticut Classical Arts Association, who offered to become part of the team and spread the word through the art community.

We needed to produce a high-quality show, so we required entrants to supply five slide photos of their work, along with a description. My wife, Carol, volunteered to join the selection committee and help view these to decide which people we would choose to participate in the show.

Hundreds of slide-bearing applications soon arrived at our door. We spent many evening hours with our slide projector reviewing the applicants' slides. We were amazed at the talent and originality that existed among these artisans and were thrilled to have chosen about 300 to join our show. We reviewed over 2,000 slides. Potters, woodworkers, glass artists, leather workers, jewelry artists and mixed media people were chosen. Patricia Yaps was able to get a couple dozen painters and sculptors to add to the show as well.

Leroy Paltrowitz, the Fair's publicist.

We offered camping on the grounds and provided cash and ribbon awards to those who were judged best in their categories. Patricia recruited a staff of judges qualified in each of the disciplines to appoint the awards.

The Fair publicist and advertising agent, Leroy Paltrowitz, prepared and circulated press releases to the local newspapers, including *The New York Times*.

Food concessions were arranged as well as staffing for the show dates, including ticket sellers and takers, security guards, reception people for the exhibitors, first aid and cleanup personnel.

We encouraged all the participants to perform demonstrations during the show to stimulate interest. We had live musical groups perform music on a center stage, giving attendees a place to sit and rest their feet. Exhibit booths could be found in several of our large buildings, a sizeable tent was erected for additional exhibit space, and several outdoor booths filled the grassy areas. Artists hung their paintings on unused animal cages and fencing as well as erecting their own tent spaces.

The weather cooperated in normal July fashion—hot and humid, but without any serious rain. The people came by the thousands. There were 25,000 the first year! The artists and craftspeople were happy with their sales and the way we ran the show and most wanted to come back again.

We ran the shows annually until the Fair closed in 1981. We made minor improvements each year and the crowds grew to 30,000 as the years went on. They were a financial success for the Fair, and we all had a good time.

The HK Collector Car Auction

In 1979, management was approached by Hyman Kusnetz of Maplewood, New Jersey, to run a collector car auction at the fairgrounds. He proposed to rent a portion of the grounds on a profit-sharing basis once the Fair's expenses were covered. He guaranteed the cost of the expenses if the income fell short. As this seemed like a no-risk situation for the Fair, we accepted.

The date was set for Saturday, June 16, with car registration and public viewing on Friday. Cars to be auctioned "included a 1913 Multiple Trophy Winner Cadillac Touring Convertible, a 1959 Ferrari 410 Superamerica, a 1955 Mercedes Benz 300SL Gullwing"[86] and a host of other exciting exotic cars.

However, that Friday presented us with a roadblock to the auto auction. The Commissioner of the Connecticut Department of Motor Vehicles declared that the auction was illegal since Mr. Kusnetz did not have a Connecticut dealer's license. A bit of panic set in and hasty calls between Mr. Kusnetz, his attorneys and the DMV finally established that, according to definition, Mr. Kusnetz was not a dealer, but a broker. He didn't own any of the cars and he wasn't buying any of the cars, but was acting as a middleman. Therefore, didn't need a license.

Although those of us who were picked to drive some of these cars across the auction block had a really good time (I got to pilot a Jaguar XK-E with a V-12 engine one lap around our racetrack), the auction wasn't well attended. Hopefully,

86 As noted in the car auction flyer.

it was successful for Mr. Kusnetz. He did make good on covering the costs, but we didn't repeat that show.

The Danbury Antique Shows

During the winter of 1979, Fair management was approached by Arnold Greenhut and William Gladstone of Westport, Connecticut, with a proposal to run antique shows at the fairgrounds. They proposed a series of two-day shows at intervals throughout the summer. The two were experienced in the antique show field and wanted to put on something big in Danbury, taking advantage of the fairgrounds' location adjacent to Interstate 84, and convenient to New York and New Jersey. Fred Fearn was interested in antiques, having visited several shows over the years, and was enthused.

We put together an agreement and booked the show dates for April 27-28, June 22-23, and September 7-8. We decided to try and attract 500 dealers and to use the Blue Parking Lot (opposite the airport runway) for the exhibition. Greenhut would handle the promotional activities, Gladstone would sign up the dealers and manage the show, and the Danbury Fair would provide the facilities, as usual.

In the antique dealer community, interest was high and the promotion garnered a lot of space in the antique and collectibles press. In readiness, Bill Gladstone and I staked out and numbered 500 exhibitor spaces in the lot and waited for the dealers to arrive.

And then the rains came...

Heavy rain began to fall Thursday night into Friday morning. It totaled nearly three inches. The parking lots were separated from the Kenosia Swamp by a small brook, which quickly overflowed and flooded all of our good work.

As the flood subsided, we quickly reallocated dealer spaces to whatever area remained reasonably dry, then imported and spread truckloads of trap rock in the road areas. The show opened, somewhat in disarray, and Saturday turned out a lot better. Most of the dealers showed up and the crowds were acceptable, considering the conditions.

The June show was initially threatened by reports of rain on Friday, but the storm clouds merely hovered and never dampened the grounds. However, a new crisis threatened our success this time.

A severe gasoline crisis had hit the country. It was created by a combination of geopolitical unrest in the Middle East, perhaps collusion among the major oil companies, and a lifting of oil price controls by President Carter, which allowed crude oil prices to sky rocket. Gas stations experienced runs on their supplies and many ran out of fuel as the panicked public strove to fill up before the prices made it unaffordable. New York State and New Jersey implemented the odd-even fill-up restrictions that had prevailed during the 1973 oil crisis. This impacted attendance at our show since the public was afraid to waste gas on weekend jaunts. Again, the turn-out was acceptable, but not what we had hoped for.

The September show was free of crisis, thankfully. The weather was great and people began to accept the fact that gasoline would be in ample supply. In the end, the shows were well-accepted by the dealers and attendance was adequate overall, but the returns to Danbury Fair, Inc., were not worth it for all the effort. Not only that, but new considerations for the Fair took the wind out of our sails.

28

THE CLOSING CURTAIN

The years had rolled on since the passing of John Leahy. Fred and I continued to make improvements in the grounds and buildings, upgraded the entertainment and the quality of the exhibits, added to the dates when events were held, and increased the income to Danbury Fair, Inc.

It seemed as if we would go on running the Fair forever.

Not a word was spoken to me about what was taking place regarding Mr. Leahy's estate. Silence was the order of the day. I knew only that Fred Fearn and Gladys Leahy would occasionally hustle off to meetings at the headquarters of the Connecticut National Bank in Bridgeport.

Usually, about two weeks before the Fair opened each year, I would move my office to the grounds in order to greet early concessionaires and assist them with their questions, problems and locations. Harold Kohler, the Fair's master carpenter and John's first cousin, would cut out nearly 500 wooden stakes for me to use in laying out the concession and ride locations. I would station myself on one of our green park benches within reach of the phone and proceed to spread out my maps and charts, labeling the stakes and binding them in bunches with an eye to efficiently hammering them into the ground.

One day, in the fall 1978, Fred Fearn came driving in, gave me a cursory wave as I sat working on my bench, and opened up the reception office at the other side of the Administration Building. Soon, the sound of a helicopter approaching the area was heard. It appeared over the grounds, hovered momentarily, then flew over the grandstand and landed on the infield. Fred hopped into his car, whipped out to the infield, and soon returned in the company of three well-dressed men. They all disappeared into the office. After a time, they emerged, were driven back to the helicopter, and flew off. Another cursory wave from Fred and out the gate he went.

Hmmm… Wonder what that was all about?

Some months went by before I finally found out. The 1978 Fair had clocked a record attendance of 378,961, and plans were being made for 1979.

Mrs. Leahy, ever a stalwart, took her place in the Treasurer's Office for every event and for every Saturday night race program. She remained in charge of her staff of assistants, ticket sellers and the receipts. By 1979, our busiest year, she was 82 years old and ever ready to do her part.

On May 16, 1979, Fred Fearn appeared in my office and said we had to go and visit Attorney Ted Gemza's office right away. So we hopped into his car and headed for Gemza's office, where the attorney and Mrs. Leahy awaited. I still had no idea what this was about.

Shortly, it was explained to me that the executors of the Estate of John W. Leahy had negotiated a two-year purchase option for the sale of the Danbury fairgrounds to the Wilmorite Corporation of Rochester, New York, for the purpose of building a regional mall on the site. As Corporate Secretary of Danbury Fair, Inc., they needed me to sign the corporate resolution authorizing this action.

Wow!

Needless to say, I was stunned. Suddenly, I knew what all the secrecy of the past few years was all about. I knew too who the people in the helicopter were. My choice at that moment was to sign the resolution or resign my position as secretary. Had I resigned, the others merely would have appointed one of themselves as secretary and had the new appointee sign it. Nor could I envision upsetting Mrs. Leahy, my elderly grandmother. I signed.

The option had already been negotiated, written and signed by the executors, Connecticut National Bank, Fred G. Fearn and Gladys Leahy for the estate, and James P. Wilmot, Chairman of the Board of Wilmorite. The option period started that very day and expired on May 15, 1981.

During this two-year period, Wilmorite was to get all the approvals and permits required from the City of Danbury, the State of Connecticut, and the Federal Government, *with the assistance of Danbury Fair, Inc., when appropriate!*

It was a massive project that encompassed building a 1.2 million sq. ft. mall and parking areas, redesigning adjacent Backus Avenue and neighboring Exit 3 of Interstate 84, controlling the waters of Kenosia Swamp and Miry Brook, and addressing concerns from the neighboring airport.

Returning to our White Street office, Fred whisked me into his office and began to explain the laws pertaining to the fiduciary responsibilities and duties of executors of an estate. He also talked about federal and state estate taxes, which were onerous for Leahy's estate. After hours of study, having considered the risks and costs of operating the Fair corporation, successful as it was, it was decided that the Fair could never compete with the income that could be produced for the estate by the outright sale of the best located, large acreage property in the area. The executors were duty-bound to make use of the estate's assets to the greatest benefit of the estate. Operating the Danbury Fair wasn't it.

So, the immediate fate of the Danbury Fair would be in limbo for two years—or less. If, and when Wilmorite exercised its option, the Fair would be discontinued and would have to remove and dispose of all of the assets that were not part of the real estate.

The very next day, May 17, headlines on the front page of *The News-Times* screamed, "Mall firm takes option on Fair," accompanied by the details as quoted by James Wilmot and Fred Fearn. The secret that had been kept from me for years was public knowledge only hours after it had been revealed to me.

SCRAM IS BORN

It wasn't long before I was contacted by Lou Funk, Jr., a high school classmate of mine (Class of '62) and a twice weekly racquetball opponent. More importantly, he was a longtime member and race driver of SNYRA, as was his father, Lou, Sr. His company, Omaha Beef, was the exclusive purveyor of meats to the Fair concessionaires, and he was very concerned.

He inquired as to how I felt about the sale of the Fair and I assured him that it was the last thing I wanted to see happen. I explained my powerlessness in the matter. He said that if I would participate in getting organized to fight the sale, he would put a meeting together. This was risky business for me, but if I could keep my name out of it, I was glad to help.

The meeting was held at his home one evening shortly after. It was also attended, as I recall, by Jack Knapp, Jim Seeley, Sr., and Jim Seeley, Jr., all men who had served SNYRA as officers throughout the years. Lou had already contacted

Attorney Jack Garamella, another classmate of ours, who, as a Danbury native, agreed to help our cause.

Garamella advised us that we had an opening gambit for our battle. The location of the fairgrounds was zoned "light industrial." Wilmorite would have to apply to the Danbury Planning Commission and the Danbury Zoning Commission to change the zone of the property to "commercial." Public hearings would be held, public testimony would be heard, and perhaps, just perhaps, the zone change could be denied.

For my part, I was to supply inside information as I gathered it, supply mailing lists of concessionaires, exhibitors, performers and Fair employees, along with seed money to help pay some early expenses and demonstrate my sincerity. The old saying, "Put your money where your mouth is," applied here.

Publicity for the proposed mall generated concern among the Danbury businessmen who owned locations downtown. Some were family businesses that had been in operation for several generations. They had noted the devastation of downtown business areas in other cities due to the construction of malls.

The proposal also generated concern among people who lived in the neighborhoods bordering the fairgrounds, especially those who had recently purchased high-end condominiums and town houses overlooking nearby Lake Kenosia. They were concerned about heavy daily traffic, the general increase in the number of people in the area, and even how the construction might affect the wetlands and the underground aquifer.

Avid race fans were outraged at the possible loss of their racetrack and fairgoers lamented the loss of a century-old tradition. Local concessionaires foresaw the loss of the extra money they made every year that helped them supplement their regular jobs, put their kids through college, and afford vacations.

It wasn't long before Garamella had organized this discord into a local community opposition effort called "Some Concerned Residents Against the Mall," or

SCRAM. The leadership of SCRAM was composed of people who represented all of the above concerns.

SCRAM made sure to contact everyone concerned with the survival of the Fair as well as those who opposed the building of the mall for other reasons. The battle was waged principally on the editorial pages of *The News-Times*. The "Opinions of the People" section was bombarded with letters from the public.

The News-Times editorial writers came out in favor of building the mall. SCRAM accused the newspaper of having an ulterior motive, that being the envisioning of millions of dollars of advertising revenue. The sheer number of "news" articles regarding the development and the developers was interpreted by many to imply the newspaper's support.

Amazingly to a number of its members, the Danbury Chamber of Commerce came out in support of the mall. A number of these members accused the board of directors of the Chamber of making this policy decision in secret, without knowledge or input from the membership. Several longtime members of the Chamber resigned in protest.

THE HEARING

Battle lines were drawn. Both sides were ready to open fire. The chosen battlefield would be the Zoning Commission meeting to take place in Danbury City Hall on March 3, 1981. The day before, on March 2, the Danbury Planning Commission, largely an advisory group, recommended denial of the zone change.

Jack Garamella recommended that I not attend the hearing as it would be highly emotional. I took that advice, as I knew that I was capable of impulsive behavior. My wife, Carol, volunteered to attend in my stead.

Due to the efforts of SCRAM and the ensuing publicity, the hearing room was packed. The climate inside was stifling and simmering emotions, kept in check for the present, could be felt in the air. The hearing opened with other petitions for zone changes that were scheduled on the agenda—a list of mundane items that the vast majority of the attendees cared nothing about. At 8:15 p.m., Zoning Commission Chairman Basil Friscia announced that a representative of the Danbury Fire Department deemed that the room was too crowded for the safety of all those in attendance. Chairman Friscia then adjourned the meeting to reopen at 9 p.m. or shortly thereafter in the gymnasium of the Rogers Park Jr. High School to take up the Wilmorite petition.

Wilmorite's testimony opened with a stream of experts from every field of endeavor that was germane to the questions that could be anticipated by the petitioner. They included testimony as to all the permits that needed to be obtained, all design considerations that needed to be addressed, details about traffic studies, details about studies regarding the nearby aquifer and flood control measures, as well as details about local shopping trends gleaned from marketing studies. Charts, maps and slideshows were presented. While much of this was certainly proper to use in presenting the developer's case, it seemed that the extreme detail of technical studies and jargon were intended to wear the crowd down in hopes they would go away.

Instead, people began to shout demands from the floor. They became unruly and seemed on the verge of becoming out of control. Chairman Friscia had to intervene several times, assuring the crowd that they would get their say.

Finally, the Wilmorite people finished their epic presentation and the meeting was opened to the floor for citizens to testify.

Many prominent Danburians strode to the microphone to speak, some in support of the mall, but most in opposition to it. Attorney Garamella identified himself as a member of SCRAM and requested time to present testimony from their own list of experts.

The hearing carried on into the night until, at 1 a.m., it was voted to adjourn and continue the meeting on March 16.

The continuance of the Zoning Commission hearing witnessed another packed house. It was estimated that about 400 people were in attendance. After taking care of a couple of other agenda items, Chairman Friscia opened the Wilmorite application testimony to "the opposition." Several local residents spoke of their concerns, not only with the sale of the Fair, but of the development itself. They were concerned with water quality, drainage issues, traffic issues, access and road changes, fiscal and business effects of the mall, including the effect the mall might have on the general quality of life in the area.

SCRAM presented its own expert, Edward Bogdan of Quality Environmental Planning Corporation, experts in environmental issues. He presented his own detailed report to the Commission and voiced his lengthy, detailed review of the state of the local aquifer and recharge areas, as well as an assessment of water quality and effects on the environment by the chemicals found in parking lot run-off.

A request to cross-examine Bogdan by Wilmorite Attorney John Shields was met with noisy and hostile objections from the attendant crowd. In view of the reaction, Friscia postponed the cross-examination until the rest of the witnesses had a chance to speak. Several members of SCRAM voiced their reasons for opposition. Former Mayor Donald Boughton testified against approval, as did Barbara McCarthy, President of the Ridgefield Chamber of Commerce.

In favor of the zone change was Martin Boucher, speaking for the Danbury Chamber of Commerce. He paraphrased the eleven-page written statement that was entered into the record as Exhibit 21. In it, the Board actually says:

> *The Board does not believe that a regional shopping mall facility would adversely affect the continued development of retail services within the various central business districts of the greater Danbury area.*

As can be witnessed today, the seeds of downtown renewal are only just beginning to germinate more than thirty years later. The Main Street business area has had a forlorn look and its recovery from the impact of the mall has been slow. Many storefronts still sit yawning emptily. Quite a few others are occupied by non-tax paying social services. Of the occupied spaces, many have seen a constant parade of business openings and closings.

After statements from approximately twenty more witnesses, sometimes affording Chairman Friscia opportunities to further warn against unruliness, the cross-examination of several of the expert witnesses began. Voluminous amounts of details, facts and figures filled the air as pointed questioning continued.

Perhaps the payoff presentation was made by Wilmorite Attorney Joseph Lane, of Bethel, Connecticut. He proceeded to read from John Leahy's will and then speculate on what Leahy might have done to continue the Fair. Lane pointed out that Leahy had failed to do any of those things. Finally, he addressed the legal duties of executors under the law.

A statement from Thomas Wilmot, President of Wilmorite Corporation (the son of James Wilmot, who had started the project and had passed away during its development), outlined the amount of effort in time, detailed studies and expenditures had been invested, "in excess of $1 million dollars." He shared that he felt they had "successfully made a case that the mall is in the best interest of the community and they would like to participate in Danbury's growth."[87]

Some further debate continued and the hearing was finally adjourned at 3 a.m. A total of eleven hours of give-and-take testimony had been heard. Now, all that could be done was to await the decision of the Zoning Commission.

On April 14, 1981, the Danbury Zoning Commission announced its decision to approve the application of Wilmorite for the zone change to CG-20, which would effectively permit the construction of the mall to ensue. The decision was a major disappointment to all who loved The Great Danbury State Fair.

On May 13, 1981, Thomas C. Wilmot had hand-delivered a letter to the Danbury Fair, Inc., Attorney Theodore Gemza, addressed to Mrs. Leahy and Mr. Fearn.

> *You are hereby notified that Wilmorite, Inc., does hereby exercise the rights granted to it under a certain Purchase Option, dated May 16, 1979, between Danbury Fair, Inc., and Wilmorite, Inc.*

---

87 Zoning Commission minutes, March 16, 1981.

This was the final stroke of the pen, just three days before the option would expire, spelling the end of the Fair. It seemed that nothing could stop it now.

In a last-ditch effort, Atty. Garamella on behalf of Karen Goodman, Chair Lady of SCRAM, Janet Gershwin, Barbara Wakefield, Gaitan LaRiviere, Carol LaRiviere, John J. Addessi (SCRAM board member), Enrico Addessi, Louis Addessi, Frank Cappiello, and Lawrence Sellman, filed suit against the Zoning Commission of the City of Danbury, Wilmorite, Inc., and Danbury Fair, Inc. The plaintiffs represented not only SCRAM, but residents of the area and downtown business owners.

The crux of the suit was that the Zoning Commission had acted illegally in that additional testimony on the zone change, not listed on the original agenda, was heard at a later hearing on April 14; that some members of the Commission had conflicts of interests that would be affected by the outcome, and a number of additional procedural transgressions.

A decision wasn't handed down until January 11, 1984. By then the fairgrounds' assets had been completely sold, dismantled and taken away. Construction of the new mall was well underway. The judge denied the validity of the suit without addressing the issues involved, by ruling that the "aggrieved" parties had no standing as defined by law. Conditions constituting "aggrievement" required ownership of property within 100 feet of the land involved in the zoning decision and that they be effected directly by the decision. None of the plaintiffs qualified in the judge's view and the suit went no further.

29

THE SHOW MUST GO ON

After the signing of the purchase option in 1979, uncertainty loomed over our plans. Whatever steps Wilmorite was taking to go forward were kept close to the vest. Even after all the resultant publicity, no further updates were revealed by them to the newspapers or to management at the Danbury Fair. As a result, we decided to move forward, continuing with the Fair, the races and the arts and crafts show. A successful season was run in 1980, with Fair attendance at 338,319.

Following the 1979 Fair, Gladys Leahy's health began to fail. She experienced a stroke that laid her low—seemingly permanently—but after a few months of rest, she snapped back to what appeared to be her previous normal condition. Recognizing that she was beginning to need help, she enlisted my sister, Susan Connally, to be Assistant Treasurer and aid in the activities of the office. Another stroke left Gladys bedridden in 1980 and, for the first time since 1946, she was absent from the Treasurer's Office.

Susan and her veteran staff carried on through the final Fairs. It was she who encouraged the treasury staff to don black armbands and emblazon the office with "Don't Mall the Fair" posters. A serious warning in the form of a letter from Wilmorite advised the Fair management and the estate's executors that these actions might result in legal retributions as they seemed to compromise the option's requirements of cooperation from Danbury Fair, Inc.

The timing of the exercise of the option allowed for the running of a final season in 1981. The stock car racing season saw the running of twenty-four Saturday night events. The Arts & Crafts Show had a record crowd.

Susan Connally, Assistant Treasurer, circa 1979.

The Fair opened as usual and the weather was perfect for the entire ten days—"John Leahy weather" prevailed. The crowds came in unprecedented numbers to enjoy the final running of the 112-year tradition. The paid attendance for the final edition of the Fair was 431,560. It was a bittersweet week for all of us who had dedicated ourselves to making the Danbury Fair an event that would live in the hearts and minds of generations of Danburians and, indeed, all the people who enjoyed it annually and made it grow. "It was the best of times, it was the worst of times..."[88]

Danbury Fair Police Lieutenant Joseph Raymond was the voice of the Fair's public address system. Daily, he called attention to events, lost children and other items of importance. The closing of the Fair hit home when his final announcement was heard ringing across the cold night air: "May I have your attention please! As usual, the tenth and last day of the Fair is upon us. Through the years I have had the privilege of closing the Fair each year. This year I must announce the falling of the curtain to a long and joyful event for the past 112 years. For the last time, I must now announce that it is now seven o'clock. All concessionaires and exhibitors...the Danbury Fair is officially closed."

Hearing the finality of his statement brought everyone up short. It was akin to watching someone lingering on their deathbed. You know what's coming, but when the actual event happens it still triggers shock and emotion.

I was accustomed, on the last night of each year's Fair, to taking a quick walk around the grounds to see how the tear-down was going. My last stop was inside the cavernous canvas-covered arena we called The Big Top. Normally, the sound of good-natured banter was heard as the exhibitors and concessionaires began to tear down and store their displays.

88 "A Tale of Two Cities," Charles Dickens.

"Good luck!"

"Have a good winter!"

"See you next year!"

Yet this last evening brought only the sound of silence. The echoing of hammers and pry bars at work resounded. Voices were muted—a sense of melancholy pervaded the air. I felt the emotion—the sadness—that had overcome all of us as the realization that there would be no "next" year sunk in. I took a deep breath, let out a sigh, turned on my heel, and headed home.

Little more than a month after the Fair closed, Gladys Leahy passed away at home on November 19, 1981. Although the state of her health did not allow her to participate in the Fair, she was certainly there in spirit. Her absence in the Treasury Office was especially felt. Even though it was a place of serious business, the staff of mostly local bankers made it a place of merriment, always providing a light atmosphere with practical jokes and friendly greetings to visitors.

Pinkerton Captain Rudy Horvath enjoys an elephant ride with Gladys Leahy.

30

AUCTIONED OFF

Our normal cleanup of the grounds took place. The Big Top came down and was stored away, while many of the figures were packed and signs were taken down. Seasonal employees were laid off. Mike Gaboardi and Pete Farnan, our two very talented sign painters, were retired. Half a dozen of our full-time staff were to work through the winter, preparing for the disposal of everything the Fair owned. The purchase option gave us several months to dispose of all such items, but the list was prodigious.

Fred Fearn engaged David Luther of Luther's Livestock Auction from nearby Wassaic, New York, to inventory and auction off everything. Luther, in turn, hired three young women to do the inventory. They worked the whole winter long, foraging through unheated barns, identifying, listing and tagging every item with a lot number to be listed in the auction program.

The giant character figures of Farmer John, Paul Bunyan, Uncle Sam, the

Indian Chief, Rip Van Winkle, the Dutch Boy and Girl, Santa Claus with his sleigh and nine reindeer (yes, Rudolph was there), toy soldiers, and the 30-foot candy canes all had to go. Dozens of other figures from fairy tale lore and images of historic figures had to go. The collection of antique carriages, wagons, farm implements, cars, trucks and tractors, as well as the one-of-a-kind locomotive, the Daniel Nason and its tender, and many other items needed new homes.

Long lists of the more mundane items were included. Tools, mowing equipment, tractors, kegs of nails, boxes of nuts and bolts, electrical supplies, lumber, unused cans of paint, plumbing supplies (including an inventory of toilets), rolls of chainlink fencing, and many dozens of signs had to go. In all, some 12,000 items were inventoried and tagged.

Luther decided that a six-day auction, involving many auctioneers, would be required. The dates were set. The last public event to be held on the historic fairgrounds would take place March 31 through April 5. A $5 admission charge was put in place in order to sort out serious buyers from the merely curious. The fee was credited toward any purchase.

March 31 opened with gloomy, drizzly skies. The wind blew with a damp, chilly breeze. It was a fitting backdrop for what seemed more like a funeral

The battle for the Fair was lost, and the Fair's figures were laid out for auction like casualties of war. April, 1982.

than an auction. The larger items, and there were many, were lined up on the midways, looking lonely and lost as if they were anxious about the uncertainty of their new homes.

Several hundred people showed up. The auctioneers with their podium and sound system rode on one of the Fair's old flatbed trucks, which, too, would be sold. Quite a few smaller items, signs and figures were purchased by local people as souvenirs. Some went to restaurants for decorative purposes. A number of the larger pieces continued their careers at an amusement park in Lake George, New York, where they could be enjoyed by future generations. The most valuable piece, the Daniel Nason steam locomotive and its tender, was purchased by the National Transportation Museum in St. Louis, Missouri, where they can still be seen today.

Santa and his reindeer up for sale with Cinderella's pumpkin coach.
Danbury Museum and Historical Society

As the auction progressed to the last day and the pickings grew slimmer, I was shocked when I saw that the windows and doors had been removed from the old caretaker's house where the Leahys had lived during the summers for many years, and where I had spent many days and nights visiting. John, Gladys and I had spent many a warm summer evening sitting on the porch in rocking chairs, eating ice cream and watching the lightning bugs.

Now the house sat with its window and door spaces yawning with blackness. The attack on the buildings continued on to the New Amsterdam Village where stained glass windows and ornate doors were stripped from their mountings and sold off.

Even the giant figures found themselves on display for the auction. April, 1982.

Finally, and mercifully, the auction came to an end.

Wilmorite closed on the purchase of the fairgrounds property on May 13, 1982. The demolition of the remains of a 112-year tradition began. It was a sad ending to "The Happiest Fair in the World." The corporation known as Danbury Fair, Inc., was dissolved on November 30, 1982.

Sam the Restaurateur was named for Sam Showan, who operated this restaurant and resembled this figure — somewhat.

EPILOGUE

The Great Danbury State Fair and John W. Leahy comprised the center of the lives of the Stetson family. John, with his unique and forceful personality, was the "godfather" of our family—and indeed to all who worked for him. He was a benevolent dictator, always making sure we had a good life—but not too good. Working for him was a sort of indentured servitude, but no one at Leahy's was ever laid off for lack of work, no one ever wanted for a steady job and paycheck—meager as it was.

I was proud to have been part of a caretaker management in running the Fair for the several years it took to settle John's estate. Fred Fearn and I put our all into it, continuing to maintain the facilities, add new events, and increase the revenue.

I was only 38 years old when my Fair management career ended. I was very angry and disappointed about the sale of the Fair—but now, more than three decades later, I can only be thankful for all those years of wonderful and unique experiences.

The demise of the Fair happened, in fact, not because of any fault of the trustees, but because John Leahy didn't follow through after it was discovered that the John W. Leahy Foundation was not the answer to preserving it. Certainly, other means to pursue the continuance of the Fair could have been found, but John was discouraged and worn down by the time, effort and expense that had been wasted setting up the Foundation.

John's experience with the Foundation served to reinforce his feelings about lawyers, which were already generally negative. Working at the fairgrounds one bright, sunny, cloudless day in perhaps the third week in May, he gathered a bunch of us "college boys" around him. He intoned, "Boys, today is a sad day. Today, Yale University is graduating four hundred new lawyers. None of them have any customers, so they'll all try to sue me!"

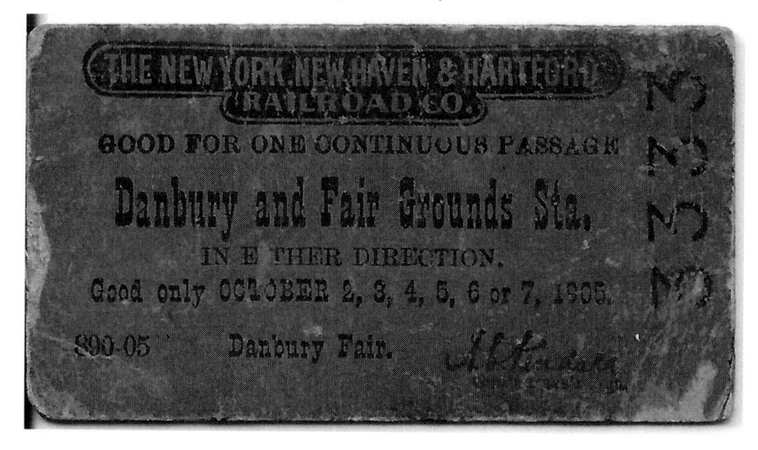

The aftermath of the sale of the Fair left a lot of other people angry besides me. To the local concessionaires, the Fair was a center of their lives, too. Locals ran most of the food concessions and it required a ton of time and effort for these family operations to prepare for participating in the Fair. These families lost not only a family tradition, but the extra income they earned. College expenses, well-earned vacations, house additions, car purchases and retirement accounts were all funded by these concessions.

The members of the Southern New York Racing Association—drivers, mechanics, pit crews, track officials—and their thousands of fans were especially upset. Stock car racing in Danbury was practically a religious cult. The club raced elsewhere a few times, but they were used to "the best," and what they found elsewhere didn't make the grade.

The hundreds of thousands of faithful fairgoers, race fans, event attendees, locals and out-of-towners all felt an empty space in their lives when the long tradition ended. For all who worked at the Fair, full- or part-time—the Danbury Fair Police, the Pinkerton Police who manned the gates and the parking lots, the night-time sanitation workers who made sure the grounds were bright and shiny each morning, the department heads and their assistants, the carnival and ride workers, and others too many to enumerate—the sale of the Fair was a personal loss.

Fred Fearn retired to enjoy life free of the huge demands of running the Leahy businesses. I took over the ownership and operation of the Leahy fuel companies, and am proud to say they are still operating today. Leahy's Fuels, Inc., will celebrate 100 years of continuous operation in 2017. Although we lost a number of customers angered by the sale of the Fair, we have continued to thrive, modernizing with the times and planning for the future.

But it ain't show biz. Long live the Great Danbury State Fair in the memories of all who enjoyed it!

JOHN H. STETSON
August 2015

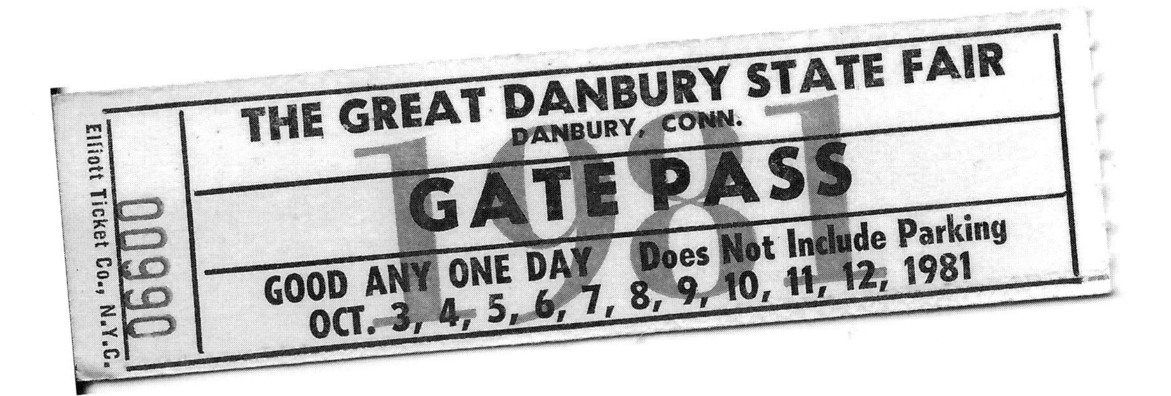

ACKNOWLEDGEMENTS

This volume was written with the help of my wife, Carol, who gave it a thorough reading, made suggestions as to tone, which I followed, and located typos, misspelings, and runtogetherwords.

I also received enthusiastic assistance from the staff of the Danbury Museum and Historical Society, namely Brigid Guertin, Michele Lee Amundsen and Diane Hassan, who loaned me photos from the museum collection, aided me with the research, and taught me to use the microfilm scanner.

C. Jarvis, Jr. who had produced the documentary video on the Fair with me, provided background information on his father, C. Irving Jarvis, who was John Leahy's Assistant General Manager for many years.

Additional thanks go to Steve Barrick of Program Dynamics, Inc., publisher of *The Life and Times of the Southern New York Racing Association*, and Lou Funk, Jr., Chairman of the SNYRA Historical Committee, for permission to use some of the photos from that memoir.

Thanks also to Tara Alemany, my publisher, for ably taking on the editing, production and publishing of this volume, a field which I knew nothing about. She was adept at preserving the odd turn of phrase, replacing the awkward word with something better, adding footnotes and captions for clarity's sake, and ensuring the logical flow of paragraph after paragraph. Many, many hours went into this effort.

Mark Gerber also played a key role in the production of this volume. He was responsible for creating the cover and overall book design. Mark partnered with Tara to perfect the physical layout, chose the font styles, selected and scanned the pictures we used from hundreds available (and took more photos himself when they were not), and designed the striking cover. I'm grateful for all his efforts on my behalf.

This seems like the proper place to thank all of those Danbury citizens who took up the cause to try and stop the Danbury Fair Mall, using their own time, money and effort. I recognize that all of their motivations were not about saving the Fair, but also reflected concerns about the environment, traffic and neighborhood atmosphere. The mall was built anyway, but all of us got a firsthand lesson in the power of big money and political influence.

Also deserving of my personal thanks are the dozens of people with whom I worked in producing the Fair, the races and the other shows we offered over the years. These people were responsible for my practical education in the fair business, the carnival business, the construction trades, and human nature, allowing for my short, but successful, show biz career.

JOHN H. STETSON

ABOUT THE AUTHORS

Gladys Stetson Leahy was born in 1897 in Poland, Maine, where she grew up with her mother in her grandparents' home, a small produce farm. A bright girl with some promise beyond her current circumstances, she graduated from Edward Little High School in Damariscotta.

Presumably under the auspices of her aunt and uncle, Harriett and Luther Spiller, she attended and graduated from Bates College with a B.A. degree in 1919. (Aunt Harriett was the Regional Superintendent of Schools and Uncle Luther was the local postmaster.)

After marrying and being deserted by her third cousin, Abner H. Stetson, and bearing his son, she embarked on a teaching career. She started that career at the Edward Little High School. Over the next several years, she changed jobs and locations frequently, seeking to increase her wages in order to support her son, John Howard, who lived with Gladys' blind mother. To further advance her position, she spent several summers gaining her M.A. degree from Boston University, which she achieved in 1928.

Her career travels eventually brought her to Danbury, Connecticut, where she began teaching at Danbury High School. She was introduced to John W. Leahy by a mutual friend in the mid-1930s. Eventually marrying him, she joined in helping her new husband with his various business ventures and was cajoled into writing a book (this book) about him and the Great Danbury State Fair.

John Howard Stetson, Jr., known to friends and family as "Jack," is her grandson and John Leahy's step-grandson. Growing up under the strong influence of both Leahys, he began his working career as an office boy at a part-time job at the age of twelve. He graduated from Danbury High School in 1962 and spent three semesters at the University of Rhode Island.

Disillusioned with college life, he left there, came home, married Carol Farwell, and spent the rest of his life as a student of the John W. Leahy College of Lifelong Learning while working his way up the ladder as an employee of the various Leahy enterprises. That culminated in his owning those businesses, which will celebrate 100 years of existence in 2017.

His many joyful years of working at the Danbury Fair prompted him to complete writing the story of its history.

INDEX

C

D

E

F

M

W

Y

Z

Danbury Museum and Historical Society

Visit us at
emeraldlakebooks.com.

24-Hour Control Gate 11-I
Administration Building 7E
Airport Gate 21-H
Amusement Area 7H
Arena Gate 9J
Art Barn 5C
Barnum Museum 6D
Better Living Center 5H
Big Horse Barn 4F
Big Top 7D
Blue Parking Lot 1A-J
Blue Ribbon Stadium 3F
Boat Ride 6C
Bus Parking Lot Orange 9A
Yellow 13I
Cattle Barn 3G
Cinderellaland 4J
Colonial Hall 6F
Connecticut Craftsmen Bldg 6F
Country Hall 5E
Daniel Nason Locomotive 4D
Electric Shop 7E
Exhibitors' Entrance 11-I
Exhibitors' Parking 12-I
First Aid Station 8J
Food Concessions 6E
Ford Museum 4E
Goats 13A
Goldtown Western Village 8D
Goldtown Music Hall 7B
Grandstand 9F
Grandstand Annex 10H
Green Parking Lot 12I
Home Show 4H
Lost & Found 8E
Maintenance Area 3D
Monkey House 11A
Motor Mall Arena 4H
New Amsterdam Gate 13-I
New Amsterdam Village 14A-J
New England Village 5B
Orange Bowl Stadium 11A
Oxen Barns 3H
Oxen Draw 3F
Petting Zoos 13A, 6F, 10I
Picnic Grove 7C-9B
Public Schools Exhibit 4C
Poultry Show 6E
Racetrack 10F
Red Parking Lot 1-15J
Restaurants 5E
Sightseeing Ride Tickets 8E
Sheep Show 5C
Swine Show 3I
Transportation Museum 4D
World of Mirth Theatre 5G
Yellow Parking Lot 15A-J

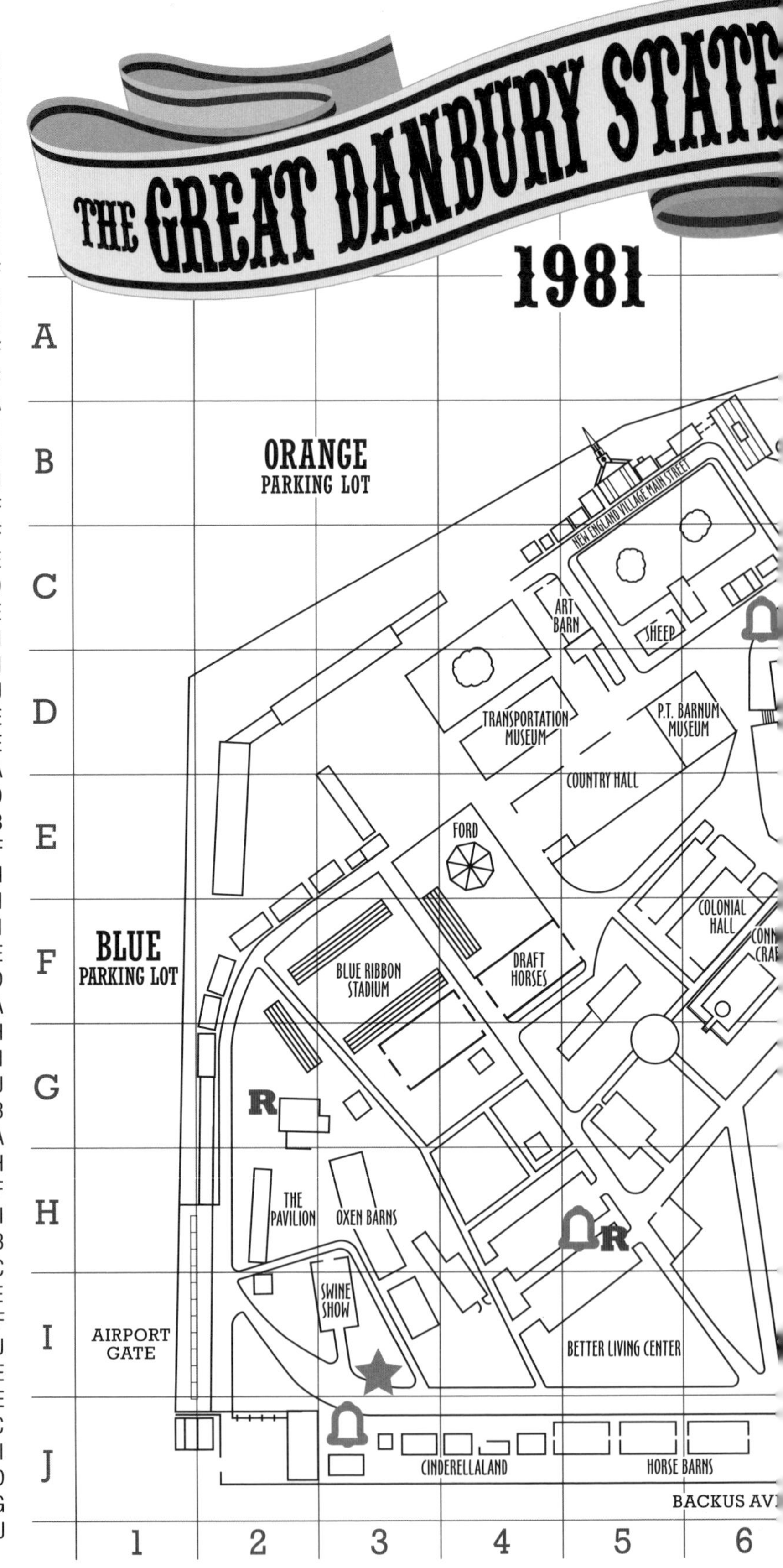